WETHERSFIELD INSTITUTE

Proceedings, 1999

SCIENCE AND EVIDENCE FOR DESIGN IN THE UNIVERSE

Science and Evidence
for Design
in the Universe

Papers Presented at a Conference
Sponsored by the Wethersfield Institute
New York City, September 25, 1999

IGNATIUS PRESS   SAN FRANCISCO

Cover design by Riz Boncan Marsella

Photo credit: Horsehead Nebula
The Royal Observatory of Edinburgh/SPL/PhotoResearchers
© Royal Observatory, Edinburgh/Anglo-Australian Observatory
Photograph by David Malin

Second printing, 2002
© 2000 Homeland Foundation
Published by Ignatius Press, San Francisco
ISBN 0-89870-809-5
Library of Congress control number 00-102374
Printed in the United States of America ⊗

WETHERSFIELD INSTITUTE
STATEMENT OF PURPOSE

The purpose of the Wethersfield Institute is to promote a clear understanding of Catholic teaching and practice and to explore the cultural and intellectual dimensions of the Catholic Faith. The Institute does so in practical ways that include seminars, colloquies and conferences especially as they pursue our goals on a scientific and scholarly level. The Institute publishes its proceedings.

It is also interested in projects that advance those subjects. The Institute usually sponsors them directly, but also joins with accredited agencies that share our interests.

CONTENTS

CONTRIBUTORS

MICHAEL J. BEHE received his Ph.D. in biochemistry from the University of Pennsylvania (1978) and is Professor of Biological Sciences at Lehigh University in Pennsylvania and a fellow of the Discovery Institute. His book, *Darwin's Black Box* (The Free Press, 1996), discusses the implications for evolutionary theory of what he calls "irreducibly complex" biochemical systems.

WILLIAM A. DEMBSKI holds a Ph.D. in mathematics from the University of Chicago, a Ph.D. in philosophy from the University of Illinois, Chicago, and an M.Div. from Princeton Theological Seminary. His most recent publications include the books *Intelligent Design* (InterVarsity, 1999) and *The Design Inference* (Cambridge University Press, 1998). He is director of the Michael Polanyi Center at Baylor University and is a fellow of the Discovery Institute.

STEPHEN C. MEYER received his Ph.D. in the History and Philosophy of Science from the University of Cambridge in 1991 for a dissertation on origin-of-life biology and the methodology of the historical sciences. He is currently Associate Professor of Philosophy at Whitworth College and the director of the Discovery Institute's Center for the Renewal of Science and Culture. He has contributed to a number of books and is currently writing a book developing a scientific theory of biological design.

FOREWORD

For more than two thousand years—many leading western thinkers, from Plato to Aquinas to Newton—argued that the natural world manifests the design of a preexistent mind or intelligence—a Creator. Yet during the late nineteenth century many scientists began to reject this idea. Charles Darwin's theory of evolution by natural selection, and other materialistic theories of the origin of life, the solar system, and the universe, portrayed nature as a self-creating and self-existent machine—one that does not show any evidence of design by a directing agency or intelligence.

Of course, even Darwinists have long acknowledged that biological organisms do "appear" to be designed. As Richard Dawkins, a leading Darwinian spokesman, has said, "Biology is the study of complicated things that give the appearance of having been designed for a purpose."[1]

Nevertheless, Darwinists have insisted that this appearance is illusory, since the mechanism of natural selection can explain the observed complexity of living things. Thus, for most of the twentieth century, science seemed to undermine the design argument and to provide little if any support for classical theistic belief.

This situation has begun to change. Over the last fifty years, discoveries, not only in biology, but also in physics, astronomy, and cosmology, suggest that life and the universe manifest signs of real, not just apparent, design. Further, many evolutionary biologists have acknowledged fundamental problems with the Darwinian mechanism as an explanation for the complexity and apparent design of living organisms. As a result of both these developments, an increasing number of scientists have rejected the idea that life and the universe merely *appear* designed. Instead, many scientists and philosophers now think the universe and life appear designed because they really were.

Many of these scientists advocate an alternative theory of biological and cosmological origins known as the *theory of intelligent design*, or, simply, design theory. Though this theory has a rich intellectual tradition, its advocates have staked out a fresh and distinctive position within the contemporary origins debate. Unlike neo-Darwinists and other evolutionary theorists, design theorists hold that intelligent causes rather than undirected natural causes best explain many features of life and the universe. Unlike many creationists, design theorists do not necessarily believe that the earth is young, neither do they base their theories upon scriptural texts. Unlike many theistic evolutionists who think design can only be seen through "the eyes of faith", design theorists believe that scientific evidence actually points to intelligent design—that intelligent design is, in their words, "empirically detectable".

In September of 1999, the Wethersfield Institute invited three leading proponents—William Dembski, Stephen Meyer, and Michael Behe—of the contemporary theory of intelligent design to Manhattan to present their case before a conference entitled "Science and Evidence for Design in the Universe". This volume makes public the essays upon which their presentations were based. An appendix includes three other essays by these same authors. These essays explore other aspects of the debate about intelligent design and respond to various scientific and philosophical criticisms of their theory.

The first essay by mathematician and probability theorist William A. Dembski provides readers with a general theory of intelligent design "detection". Dembski notes, first, that good reasoning often leads people to infer the activity of intelligent agents from the effects that they leave behind. He uses a variety of examples—from archeology, cryptography, and fraud detection—to show that humans routinely infer that intelligence played a role in the origin of certain kinds of artifacts

or events. Dembski's work shows why by making explicit the
criteria that we use to make such inferences. He argues that
whenever we observe events that are "highly improbable" (i.e.,
complex) and "specified", we make (well justified) design inferences. His work in effect establishes a scientific method of
detecting the activity of intelligence.

In chapter two, philosopher of science Stephen Meyer uses
Dembski's method to examine evidence from the natural world.
He first examines the so-called "fine-tuning" of the laws of
physics. He shows that this feature of the universe exemplifies
Dembski's criteria of design. For this and other reasons, he
argues that intelligent design best explains the origin of the
fine-tuning evidence. He then makes a similar argument about
the origin of the information necessary to build a living cell.
He notes that studies of the genetic molecule DNA reveal that
it functions in much the same way as a computer code or
written text. Accordingly, he shows that DNA possesses both
the complexity and specificity of function that, according to
Dembski's theory, indicate intelligent design. He concludes
that the information content of DNA—like the information
in a computer program or an ancient scroll—had an intelligent
source.

In chapter three, biochemist Michael Behe (author of *Darwin's Black Box* [Free Press, 1996]) describes other evidences
of design—the complex motors that reside in cells and the
Rube Goldberg-like "vision cascade" responsible for light sen-
sitivity in the eye. He characterizes these biochemical systems
as "irreducibly complex" because they require many separate
proteins parts working together in order to function. Behe argues that such systems are unlikely to have originated by the
neo-Darwinian mechanism. Natural selection can only act on
systems that perform functions that help organisms survive.
But "irreducibly complex" systems have no function at all unless all parts in the system are present. Yet without the aid of
natural selection the odds against such systems arising on their

own are prohibitive. For this reason, Behe elects intelligent design as a better explanation than neo-Darwinism (or chance) for the origin of irreducibly complex systems and motors.

In the next essay (the first of the appendix), Michael Behe responds to scientific criticism of his design argument. Though, to date, most published criticism of Behe's argument has been methodological or philosophical in character (see below), some has been scientific. In this last chapter, Behe responds directly to his staunchest scientific critics, including biologist Kenneth Miller of Brown University. In the process, he brings readers up-to-date on the status of the contemporary scientific argument about intelligent design in irreducibly complex systems.

Chapter 5 in the appendix returns specifically to the question of design. It addresses an objection that has been raised repeatedly against scientific or empirical arguments for intelligent design. Critical reviewers of Michael Behe's *Darwin's Black Box*, for example, generally have not objected to Behe's argument on scientific grounds. Instead, critics have mostly objected to Behe's work on methodological grounds. Behe's critics have often claimed that to infer an intelligent cause (as Behe does) violates the "rules of science" or "goes beyond science". Scientific explanations, they assert, must limit themselves to strictly naturalistic causes. Stephen Meyer's essay "The Scientific Status of Intelligent Design" examines this objection. He shows that no clear methodological justification exists for defining science in this way, and that instead, attempts to do so inevitably limit the truth-seeking function of science.

In the concluding chapter, we situate the present debate about design within the larger discussion of the relationship between science and religion. In this chapter, William Dembski and Stephen Meyer clarify the way in which scientific evidence might lend support to theistic belief. They acknowledge that deductive arguments for the existence of a transcendent personal God often fail to produce the certainty that they promise. Nevertheless, they argue that a lack of deductive certainty does not leave scepticism or blind faith as the only alternatives. Physical evidence from nature (such as current astronomical evidence for a finite universe) might justify theism as "an inference to the best explanation", even if such evidence cannot provide the basis for indubitable proofs of God's existence.

[1] Richard Dawkins, *The Blind Watchmaker: Why the Evidence of Evolution Reveals a Universe without Design* (London: Penguin Books, 1987).

William A. Dembski

THE THIRD MODE OF EXPLANATION: DETECTING EVIDENCE OF INTELLIGENT DESIGN IN THE SCIENCES

1. Introduction

In our workaday lives we find it important to distinguish between three modes of explanation: necessity, chance, and design. Did she fall, or was she pushed? And if she fell, was her fall accidental or unavoidable? To say she was pushed is to attribute her plunge to design. To say her fall was accidental or unavoidable is to attribute her plunge respectively to chance or necessity. More generally, given an event, object, or structure, we want to know: Did it have to happen? Did it happen by accident? Did an intelligent agent cause it to happen? In other words, did it happen by necessity, chance, or design?

At this level of analysis, necessity, chance, and design remain pretheoretical and therefore inadequate for constructing a scientific theory of design. It is therefore fair to ask whether there is a principled way to distinguish these modes of explanation. Philosophers and scientists have disagreed not only about how to distinguish these modes of explanation but also about their very legitimacy. The Epicureans, for instance, gave pride of place to chance. The Stoics, on the other hand, emphasized necessity and design but rejected chance. In the Middle Ages Moses Maimonides contended with the Islamic interpreters of Aristotle who viewed the heavens as, in Maimonides' words, "the necessary result of natural laws".[1] Where the Islamic philosophers saw necessity, Maimonides saw design.

17

In arguing for design in his *Guide for the Perplexed*, Maimonides looked to the irregular distribution of stars in the heavens. For him that irregularity demonstrated contingency (that is, an event that happened but did not have to happen and therefore was not necessary). But was that contingency the result of chance or design? Neither Maimonides nor the Islamic interpreters of Aristotle had any use for Epicurus and his views on chance. For them chance could never be fundamental but was at best a placeholder for ignorance. Thus for Maimonides and his Islamic colleagues, the question was whether a principled distinction could be drawn between necessity and design. The Islamic philosophers, intent on keeping Aristotle pure of theology, said no. Maimonides, arguing from observed contingency in nature, said yes. His argument focused on the distribution of stars in the night sky:

> What determined that the one small part [of the night sky] should have ten stars, and the other portion should be without any star? . . . The answer to [this] and similar questions is very difficult and almost impossible, if we assume that all emanates from God as the necessary result of certain permanent laws, as Aristotle holds. But if we assume that all this is the result of design, there is nothing strange or improbable; the only question to be asked is this: What is the cause of this design? The answer to this question is that all this has been made for a certain purpose, though we do not know it; there is nothing that is done in vain, or by chance. . . . How, then, can any reasonable person imagine that the position, magnitude, and number of the stars, or the various courses of their spheres, are purposeless, or the result of chance? There is no doubt that every one of these things is . . . in accordance with a certain design; and it is extremely improbable that these things should be the necessary result of natural laws, and not that of design.[2]

Modern science has also struggled with how to distinguish between necessity, chance, and design. Newtonian mechanics, construed as a set of deterministic physical laws, seemed only to permit necessity. Nonetheless, in the General Scholium to his *Principia*, Newton claimed that the stability of the planetary system depended not only on the regular action of the universal law of gravitation but also on the precise initial positioning of the planets and comets in relation to the sun. As he explained:

> Though these bodies may, indeed, persevere in their orbits by the mere laws of gravity, yet they could by no means have at first derived the regular position of the orbits themselves from those laws. . . . [Thus] this most beautiful system of the sun, planets, and comets, could only proceed from the counsel and dominion of an intelligent and powerful being.[3]

Like Maimonides, Newton saw both necessity and design as legitimate explanations but gave short shrift to chance.

Newton published his *Principia* in the seventeenth century. Yet by the nineteenth century necessity was still in, chance was still out, but design had lost much of its appeal. When asked by Napoleon where God fit into his equations of celestial mechanics, Laplace famously replied, "Sire, I have no need of that hypothesis." In place of a designing intelligence that precisely positioned the heavenly bodies, Laplace proposed his nebular hypothesis, which accounted for the origin of the solar system strictly through natural gravitational forces.[4]

Since Laplace's day, science has largely dispensed with design. Certainly Darwin played a crucial role here by eliminating design from biology. Yet at the same time science was dispensing with design, it was also dispensing with Laplace's vision of a deterministic universe (recall Laplace's famous demon who could predict the future and retrodict the past with perfect precision provided that present positions and momenta of particles were fully known).[5] With the rise of statistical mechanics and then quantum mechanics, the role of chance in physics came to be regarded as ineliminable. Especially convincing here has been the failure of the Bell inequality.[6] Consequently, a deterministic, necessitarian universe has given way to a stochastic universe in which chance and necessity are both regarded as fundamental modes of scientific explanation, neither being reducible to the other. To sum up, contemporary science allows a principled distinction between necessity and chance but repudiates design as a possible explanation for natural phenomena.

2. Rehabilitating Design

But was science right to repudiate design? In *The Design Inference* I argue that design is a legitimate and fundamental mode of scientific explanation, on a par with chance and necessity.[7] In arguing this claim, however, I want to avoid prejudging the implications of design for science. In particular, it is not my aim to guarantee creationism. Design, as I develop it, cuts both ways and might just as well be used to defeat creationism by clarifying the superfluity of design in biology. My aim is not to find design in any one place but to open up possibilities for finding design as well as for shutting it down.

My aim, then, is to rehabilitate design as a mode of scientific explanation. Given that aim, it will help to review why design was removed from science in the first place. Design, in the form of Aristotle's formal and final causes, had after all once occupied a perfectly legitimate role within natural philosophy, or what we now call science. With the rise of modern science, however, these causes fell into disrepute.

We can see how this happened by considering Francis Bacon. Bacon, a contemporary of Galileo and Kepler, though himself not a scientist, was a terrific propagandist for science. Bacon concerned himself much about the proper conduct of science, providing detailed canons for experimental observation, recording of data, and inferences from data. What interests us here, however, is what he did with Aristotle's four causes. For Aristotle, to understand any phenomenon properly, one had to understand its four causes, namely, its material, efficient, formal, and final cause.[8]

A standard example philosophers use to illustrate Aristotle's four causes is to consider a statue—say Michelangelo's *David*.

The material cause is what it is made of—marble. The efficient cause is the immediate activity that produced the statue —Michelangelo's actual chipping away at a marble slab with hammer and chisel. The formal cause is its structure—it is a representation of David and not some random chunk of marble. And finally, the final cause is its purpose—presumably, to beautify some Florentine palace.

Two points about Aristotle's causes are relevant to this discussion. First, Aristotle gave equal weight to all four causes. In particular, Aristotle would have regarded any inquiry that omitted one of his causes as fundamentally deficient. Second, Bacon adamantly opposed including formal and final causes within science (see his *Advancement of Learning*).[9] For Bacon, formal and final causes belonged to metaphysics and not to science. Science, according to Bacon, needed to limit itself to material and efficient causes, thereby freeing science from the sterility that inevitably results when science and metaphysics are conflated. This was Bacon's line, and he argued it forcefully.

We see Bacon's line championed in our own day by atheists and theists alike. In *Chance and Necessity*, biologist and Nobel laureate Jacques Monod argued that chance and necessity alone suffice to account for every aspect of the universe. Now whatever else we might want to say about chance and necessity, they provide at best a reductive account of Aristotle's formal causes and leave no room whatever for Aristotle's final causes. Indeed, Monod explicitly denies any place for purpose within science.[10]

Monod was an outspoken atheist. Nevertheless, as outspoken a theist as Stanley Jaki will agree with Monod about this aspect of science. Jaki is as theologically conservative a historian of science and Catholic priest as one is likely to find. Yet in his published work he explicitly states that purpose is a purely metaphysical notion and cannot legitimately be included within science. Jaki's exclusion of purpose, and more generally design, from science has practical implications. For instance, it leads him to regard Michael Behe's project of inferring biological design from irreducibly complex biochemical systems as misguided.[11]

Now I do not want to give the impression that I am advocating a return to Aristotle's theory of causation. There are problems with Aristotle's theory, and it needed to be replaced. My concern, however, is with what replaced it. By limiting scientific inquiry to material and efficient causes, which are of course perfectly compatible with chance and necessity, Bacon championed a view of science that could only end up excluding design.

But suppose we lay aside a priori prohibitions against design. In that case, what is wrong with explaining something as designed by an intelligent agent? Certainly there are many everyday occurrences which we explain by appealing to design. Moreover, in our workaday lives it is absolutely crucial to distinguish accident from design. We demand answers to such questions as, Did she fall, or was she pushed? Did someone die accidentally or commit suicide? Was this song conceived independently, or was it plagiarized? Did someone just get lucky on the stock market, or was there insider trading?

Not only do we demand answers to such questions, but entire industries are devoted to drawing the distinction between accident and design. Here we can include forensic science, intellectual property law, insurance claims investigation, cryptography, and random number generation—to name but a few. Science itself needs to draw this distinction to keep itself honest. As a January 1998 issue of *Science* made clear, plagiarism and data falsification are far more common in science than we would like to admit.[12] What keeps these abuses in check is our ability to detect them.

If design is so readily detectable outside science, and if its detectability is one of the key factors keeping scientists honest, why should design be barred from the actual content of science? There is a worry here. The worry is that when we leave the constricted domain of human artifacts and enter the unbounded domain of natural objects, the distinction between design and nondesign cannot be reliably drawn. Consider, for instance, the following remark by Darwin in the concluding chapter of his *Origin of Species*:

> Several eminent naturalists have of late published their belief that a multitude of reputed species in each genus are not real species; but that other species are real, that is, have been independently created. . . . Nevertheless they do not pretend that they can define, or even conjecture, which are the created forms of life, and which are those produced by secondary laws. They admit variation as a vera causa in one case, they arbitrarily reject it in another, without assigning any distinction in the two cases.[13]

Darwin is here criticizing fellow biologists who claim that some species result from purely natural processes but that other species are specially created. According to Darwin, these biologists failed to provide any objective method for distinguishing between those forms of life that were specially created and those that resulted from natural processes (or what Darwin calls "secondary laws"). Yet without such a method for distinguishing the two, how can we be sure that our ascriptions of design hold water? It is this worry of falsely ascribing something to design (here construed as creation) only to have it overturned later that has prevented design from entering science proper.

This worry, though perhaps justified in the past, can no longer be sustained. There does in fact exist a rigorous criterion for discriminating intelligently from unintelligently caused objects. Many special sciences already use this criterion, though in a pretheoretic form (for example, forensic science, artificial intelligence, cryptography, archeology, and the Search for Extraterrestrial Intelligence). In *The Design Inference* I identify and make precise this criterion. I call it the *complexity-specification criterion*. When intelligent agents act, they leave behind a characteristic trademark or signature—what I define as *specified complexity*. The complexity-specification criterion detects design by identifying this trademark of designed objects.[14]

3. The Complexity-Specification Criterion

A detailed explication and justification of the complexity-specification criterion is technical and can be found in *The Design Inference*. Nevertheless, the basic idea is straightforward and easily illustrated. Consider how the radio astronomers in the movie *Contact* detected an extraterrestrial intelligence. This movie, based on a novel by Carl Sagan, was an enjoyable piece of propaganda for the SETI research program—the Search for Extraterrestrial Intelligence. To make the movie interesting, the SETI researchers in *Contact* actually did find an extraterrestrial intelligence (the *non*-fictional SETI program has yet to be so lucky).

How, then, did the SETI researchers in *Contact* convince themselves that they had found an extraterrestrial intelligence? To increase their chances of finding an extraterrestrial intelligence, SETI researchers monitor millions of radio signals from outer space. Many natural objects in space produce radio waves (for example, pulsars). Looking for signs of design among all these naturally produced radio signals is like looking for a needle in a haystack. To sift through the haystack, SETI researchers run the signals they monitor through computers programmed with pattern-matchers. So long as a signal does not match one of the preset patterns, it will pass through the pattern-matching sieve (even if it has an intelligent source). If, on the other hand, it does match one of these patterns, then, depending on the pattern matched, the SETI researchers may have cause for celebration.

The SETI researchers in *Contact* did find a signal worthy of celebration, namely, the following:

```
110110111101111111011111111111011111111111110
111111111111111101111111111111111111011111111
111111111111101111111111111111111111111111011
111111111111111111111111111011111111111111111
```

```
IIIIIIIIIIIIIIIIIIIIIIIOIIIIIIIIIIIIIIIIIIIIIIIIIIIII
IIIIIIIIIIIIIIIIOIIIIIIIIIIIIIIIIIIIIIIIIIIIIIIIIIIII
IIIIIIIIIIIIIOIIIIIIIIIIIIIIIIIIIIIIIIIIIIIIIIIIIIIII
IIIIIIIIIIIIIOIIIIIIIIIIIIIIIIIIIIIIIIIIIIIIIIIIIIIII
IIIIIIIIIIIIIIIIIIIOIIIIIIIIIIIIIIIIIIIIIIIIIIIIIIIII
IIIIIIIIIIIIIIIIIIIIIIIIIIIIIIIIIIIIIOIIIIIIIIIIIIIII
IIIIIIIIIIIIIIIIIIIIIIIIIIIIIIIIIIIIIIIIIIIIIIIIIIIII
IIIIIIIIOIIIIIIIIIIIIIIIIIIIIIIIIIIIIIIIIIIIIIIIIIIII
IIIIIIIIIIIIIIIIIIIIIIIIIIIIIIIIIIIOIIIIIIIIIIIIIIIII
IIIIIIIIIIIIIIIIIIIIIIIIIIIIIIIIIIIIIIIIIIIIIIIIIIIII
IIIIIIIIIIIIIOIIIIIIIIIIIIIIIIIIIIIIIIIIIIIIIIIIIIIII
IIIIIIIIIIIIIIIIIIIIIIIIIIIIIIIIIIIIIIIIIIIIIIIIIIIII
OIIIIIIIIIIIIIIIIIIIIIIIIIIIIIIIIIIIIIIIIIIIIIIIIIIII
IIIIIIIIIIIIIIIIIIIIIIIIIIIIIIIIIIIIIOIIIIIIIIIIIIIII
IIIIIIIIIIIIIIIIIIIIIIIIIIIIIIIIIIIIIIIIIIIIIIIIIIIII
IIIIIIIIIIIIIIIIIIIIIIIIIIIIIIIIIIIIIOIIIIIIIIIIIIIII
IIIIIIIIIIIIIIIIIIIIIIIIIIIIIIIIIIIIIIIIIIIIIIIIIIIII
IIIIIIIIIIIIIIIIIIIIIIIIIIIIIIIIIIIIIIIOIIIIIIIIIIIII
IIIIIIIIIIIIIIIIIIIIIIIIIIIIIIIIIIIIIIIIIIIIIIIIIIIII
IIIIIIIIIIIIIIIIIIIIIIIIIIIIIIIIIIIIIIIIIIIIIIIIIIII
```

The SETI researchers in *Contact* received this signal as a sequence of 1126 beats and pauses, where 1's correspond to beats and 0's to pauses. This sequence represents the prime numbers from 2 to 101, where a given prime number is represented by the corresponding number of beats (that is, 1's), and the individual prime numbers are separated by pauses (that is, 0's). The SETI researchers in *Contact* took this signal as decisive confirmation of an extraterrestrial intelligence.

What about this signal indicates design? Whenever we infer design, we must establish three things: *contingency*, *complexity*, and *specification*. Contingency, by which we mean that an event was one of several possibilities, ensures that the object is not the result of an automatic and hence unintelligent process. Complexity ensures that the object is not so simple that it can readily be explained by chance. Finally, specification ensures that the object exhibits the type of pattern characteristic of intelligence. Let us examine these three requirements more closely.

In practice, to establish that an object, event, or structure is contingent, one must show that it is not the result of a natural law (or algorithm). For example, a crystal of salt results from forces of chemical necessity that can be described by the laws of chemistry. By contrast, a setting of silverware is not. No physical or chemical laws dictate that the fork must be on the left and the knife and spoon on the right. The place setting of silverware is therefore contingent, whereas the structure of the crystal is the result of physical necessity. Michael Polanyi and Timothy Lenoir have both described a method of establishing contingency.[15] The method applies quite generally: the position of Scrabble pieces on a Scrabble board is irreducible to the natural laws governing the motion of Scrabble pieces; the configuration of ink on a sheet of paper is irreducible to the physics and chemistry of paper and ink; the sequencing of DNA bases is irreducible to the bonding affinities between the bases; and so on. In the case of the radio signal in *Contact*, the pattern of o's and 1's forming a sequence of prime numbers is irreducible to the laws of physics that govern the transmission of radio signals. We therefore regard the sequence as contingent.

To see next why complexity is crucial for inferring design, consider the following sequence of bits:

$$110111011111$$

These are the first twelve bits in the previous sequence representing the prime numbers 2, 3, and 5 respectively. Now it is a sure bet that no SETI researcher, if confronted with this twelve-bit sequence, is going to contact the science editor at the *New York Times*, hold a press conference, and announce that an extraterrestrial intelligence has been discovered. No headline is going to read, "Aliens Master First Three Prime Numbers!" The problem is that this sequence is much too short (and thus too simple) to establish that an extraterrestrial intelligence with knowledge of prime numbers produced it. A randomly beating radio source might by chance just happen to output this sequence. A sequence of 1126 bits representing the prime numbers from 2 to 101, however, is a different story. Here the sequence is sufficiently long (and therefore sufficiently complex) that only an extraterrestrial intelligence could have produced it.

Complexity as I am describing it here is a form of probability. Later in this paper I will require a more general conception of complexity. But for now complexity as a form of probability is all we need. To see the connection between complexity and probability, consider a combination lock. The more possible combinations of the lock, the more complex the mechanism and correspondingly the more improbable that the mechanism can be opened by chance. A combination lock whose dial is numbered from 0 to 39 and which must be turned in three alternating directions will have 64,000 (= 40 × 40 × 40) possible combinations and thus a 1/64,000 probability of being opened by chance. A more complicated combination lock whose dial is numbered from 0 to 99 and which must be turned in five alternating directions will have 10,000,000,000 (= 100 × 100 × 100 × 100 × 100) possible combinations and thus a 1/10,000,000,000 probability of being opened by chance. Complexity and probability therefore vary inversely: the greater the complexity, the smaller the probability. Thus to determine whether something is sufficiently complex to warrant a design inference is to determine whether it has sufficiently small probability.

Even so, complexity (or improbability) is not enough to eliminate chance and establish design. If I flip a coin 1000 times, I will participate in a highly complex (that is, highly improbable) event. Indeed, the sequence I end up flipping will be one in a trillion trillion trillion . . . , where the ellipsis needs twenty-two more "trillions". This sequence of coin tosses will not, however, trigger a design inference. Though complex, this

sequence will not exhibit a suitable pattern. Contrast this with
the previous sequence representing the prime numbers from 2
to 101. Not only is this sequence complex, but it also embod-
ies a suitable pattern. The SETI researcher who in the movie
Contact discovered this sequence put it this way: "This isn't
noise, this has structure."

What is a *suitable* pattern for inferring design? Not just any
pattern will do. Some patterns can legitimately be employed
to infer design whereas others cannot. The way in which we
make this distinction is easily illustrated. Consider the case of
an archer. Suppose an archer stands fifty meters from a large
wall with bow and arrow in hand. The wall, let us say, is
sufficiently large that the archer cannot help but hit it. Now
suppose each time the archer shoots an arrow at the wall, the
archer paints a target around the arrow so that the arrow sits
squarely in the bull's-eye. What can we conclude from this
scenario? Absolutely nothing about the archer's ability as an
archer. Yes, there is a pattern being matched; but it is a pattern
fixed only after the arrow has been shot. Thus the pattern is
contrived, or what I call "fabricated" (see below).

But suppose instead the archer paints a fixed target on the
wall and then shoots at it. Suppose the archer shoots a hundred arrows and each time hits a perfect bull's-eye. What can
be concluded from this second scenario? Confronted with this
second scenario, we are obligated to infer that here is a world-
class archer, one whose shots can legitimately be attributed,
not to luck, but rather to the archer's skill and mastery. Skill
and mastery are, of course, types of design.

The type of pattern where an archer fixes a target first and
then shoots at it is common to statistics, where it is known as
setting a rejection region prior to an experiment. In statistics,
if the outcome of an experiment falls within a rejection region,
the chance hypothesis supposedly responsible for the outcome
is rejected. The reason for setting a rejection region prior to
an experiment is to forestall what statisticians call "data snoop-
ing", or "cherry picking". Just about any data set will contain

strange and improbable patterns if we look hard enough. By forcing experimenters to set their rejection regions prior to an experiment, the statistician protects the experiment from spurious patterns that could just as well result from chance.

Now a little reflection makes clear that a pattern need not be given prior to an event to eliminate chance and implicate design. Consider the following cipher text:

nfuijolt ju jt mjlf b xfbtfm

Initially this looks like a random sequence of letters and spaces —initially one detects no pattern on the basis of which to reject chance and infer design.

But suppose next that someone comes along and tells you to treat this sequence as a Caesar cipher, moving each letter one notch down the alphabet. Now the sequence reads,

methinks it is like a weasel

Even though the pattern (in this case, the decrypted text) is given after the fact, it still is the right sort of pattern for eliminating chance and inferring design. In contrast to statistics, which always identifies its patterns before an experiment is performed, cryptanalysis must discover its patterns after the fact. In both instances, however, the patterns are suitable for inferring design.

Although in the example of the archer, the pattern (the target) is established before the event (that is, before the arrow is shot) that conforms to it, and in the example of the "methinks it is like a weasel" sequence, the pattern is only recognized after the fact, both patterns clearly indicate prior design by an intelligence. But why? What is it about these two patterns that indicates the activity of an intelligence, whereas other patterns (like the target drawn around the arrow after it is shot) do not? The key concept is that of "independence". I define a specification as a match between an event and an independently given pattern. Events that are both highly complex and specified (that is, that match an independently given pattern) indicate design.

In the first case, where the archer hits a target that exists prior to his shooting the arrow, the pattern is clearly independent of the event. The pattern existed, and was known to exist, before the event occurred. When the arrow hits the target, an event (the arrow shot) conforms to an independently given pattern (the target). In the other case, where the archer draws the pattern around the arrow, the event does not conform to an independently existing pattern (the target). Instead, the pattern (the target) was made to conform to (or was derived from) the event in question. This type of nonindependent pattern I call a fabrication. Fabrications do not indicate anything about whether the event in question was designed.[16]

In the third case of the "methinks it is like a weasel" sequence, the pattern (a meaningful string of English characters) is recognized after the fact but still indicates design. Why? The answer is, again, that the pattern is independent of the event in question. In this case the event in question (the cipher text) conforms to a set of preexisting conventions of English vocabulary and grammar, indeed, to a specific sentence from a Shakespeare play. The pattern does not exist independently of the reception of the text (the event in question), even though we may only recognize the pattern later after some reflection. Indeed, upon analyzing the text we recognize that the text conforms to the independently existing conventions of English vocabulary and grammar. Thus, the pattern imbedded in the cipher text is independent of the event of our reading and analyzing it. For this reason, we have a specification, not a fabrication, and hence, evidence (in conjunction with the complexity of the sequence) for intelligent design. Technically trained readers will want to know that the distinction between a specification and a fabrication (illustrated and described above) can be justified rigorously by employing the notion of conditional independence.[17]

Patterns thus divide into two types, those that in the presence of complexity warrant a design inference and those that despite the presence of complexity do not warrant a design inference. The first type of pattern I call a *specification*, the second a *fabrication*. Specifications are the non–ad hoc patterns that can legitimately be used to eliminate chance and warrant a design inference. In contrast, fabrications are the ad hoc patterns that cannot legitimately be used to warrant a design inference. This distinction between specifications and fabrications can be made with full statistical rigor.[18]

To sum up, the complexity-specification criterion detects design by establishing three things: contingency, complexity, and specification. When called to explain an event, object, or structure, we have a decision to make—are we going to attribute it to *necessity*, *chance*, or *design*? According to the complexity-specification criterion, to answer this question is to answer three simpler questions: Is it contingent? Is it complex? Is it specified? Consequently, the complexity-specification criterion can be represented as a flow chart with three decision nodes. I call this flow chart the Explanatory Filter. [See figure on p. 32.]

5. False Negatives and False Positives

As with any criterion, we need to make sure that the judgments of the complexity-specification criterion agree with reality. Consider medical tests. Any medical test is a criterion. A perfectly reliable medical test would detect the presence of a disease whenever it is indeed present and fail to detect the disease whenever it is absent. Unfortunately, no medical test is perfectly reliable, and so the best we can do is keep the proportion of false positives and false negatives as low as possible.

All criteria, and not just medical tests, face the problem of false positives and false negatives. A criterion attempts to classify individuals with respect to a target group (in the case of medical tests, those who have a certain disease). When the criterion places in the target group an individual who should not be there, it commits a false positive. Alternatively, when the criterion fails to place in the target group an individual who should be there, it commits a false negative.

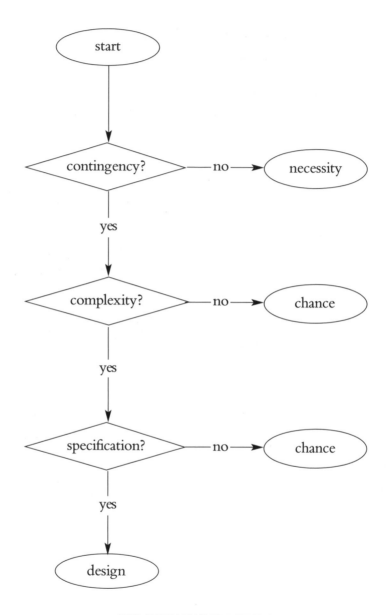

THE EXPLANATORY FILTER

Let us now apply these observations to the complexity-specification criterion. This criterion purports to detect design. Is it a reliable criterion? The target group for this criterion comprises all things intelligently caused. How accurate is this criterion at correctly assigning things to this target group and correctly omitting things from it? The things we are trying to explain have causal stories. In some of those causal stories intelligent causation is indispensable, whereas in others it is dispensable. An inkblot can be explained without appealing to intelligent causation; ink arranged to form meaningful text cannot. When the complexity-specification criterion assigns something to the target group, can we be confident that it actually is intelligently caused? If not, we have a problem with false positives. On the other hand, when this criterion fails to assign something to the target group, can we be confident that no intelligent cause underlies it? If not, we have a problem with false negatives.

Consider first the problem of false negatives. When the complexity-specification criterion fails to detect design in a thing, can we be sure no intelligent cause underlies it? The answer is No. For determining that something is not designed, this criterion is not reliable. False negatives are a problem for it. This problem of false negatives, however, is endemic to detecting intelligent causes.

One difficulty is that intelligent causes can mimic necessity and chance, thereby rendering their actions indistinguishable from such unintelligent causes. A bottle of ink may fall off a cupboard and spill onto a sheet of paper. Alternatively, a human agent may deliberately take a bottle of ink and pour it over a sheet of paper. The resulting inkblot may look identical in both instances but, in the one case, results by chance, in the other by design.

Another difficulty is that detecting intelligent causes requires background knowledge on our part. It takes an intelligent cause to know an intelligent cause. But if we do not know enough, we will miss it. Consider a spy listening in on a communication channel whose messages are encrypted. Unless the spy knows how to break the cryptosystem used by the parties on whom he is eavesdropping, any messages passing the communication channel will be unintelligible and might in fact be meaningless.

The problem of false negatives therefore arises either when an intelligent agent has acted (whether consciously or unconsciously) to conceal his actions or when an intelligent agent in trying to detect design has insufficient background knowledge to determine whether design actually is present. Detectives face this problem all the time. A detective confronted with a murder needs first to determine whether a murder has indeed been committed. If the murderer was clever and made it appear that the victim died by accident, then the detective will mistake the murder for an accident. So too, if the detective is stupid and misses certain obvious clues, the detective will mistake the murder for an accident. In mistaking a murder for an accident, the detective commits a false negative. Contrast this, however, with a detective facing a murderer intent on revenge and who wants to leave no doubt that the victim was intended to die. In that case the problem of false negatives is unlikely to arise (though we can imagine an incredibly stupid detective, like Chief Inspector Clouseau, mistaking a rather obvious murder for an accident).

Intelligent causes can do things that unintelligent causes cannot and can make their actions evident. When for whatever reason an intelligent cause fails to make its actions evident, we may miss it. But when an intelligent cause succeeds in making its actions evident, we take notice. This is why false negatives do not invalidate the complexity-specification criterion. This criterion is fully capable of detecting intelligent causes intent on making their presence evident. Masters of stealth intent on concealing their actions may successfully evade the criterion. But masters of self-promotion intent on making sure their intellectual property gets properly attributed find in the complexity-specification criterion a ready friend.

And this brings us to the problem of false positives. Even

though specified complexity is not a reliable criterion for *eliminating* design, it is, I shall argue, a reliable criterion for *detecting* design. The complexity-specification criterion is a net. Things that are designed will occasionally slip past the net. We would prefer that the net catch more than it does, omitting nothing due to design. But given the ability of design to mimic unintelligent causes and the possibility that, due to ignorance, we will pass over things that are designed, this problem cannot be remedied. Nevertheless, we want to be very sure that whatever the net does catch includes only what we intend it to catch, to wit, things that are designed. If this is the case, we can have confidence that whatever the complexity-specification criterion attributes to design is indeed designed. On the other hand, if things end up in the net that are not designed, the criterion will be worthless.

I want, then, to argue that specified complexity is a reliable criterion for detecting design. Alternatively, I want to argue that the complexity-specification criterion successfully avoids false positives. Thus, whenever this criterion attributes design, it does so correctly. Let us now see why this is the case. I offer two arguments. The first is a straightforward inductive argument: in every instance where the complexity-specification criterion attributes design, and where the underlying causal story is known (that is, where we are not just dealing with circumstantial evidence, but where, as it were, the video camera is running and any putative designer would be caught redhanded), it turns out design actually is present; therefore, design actually is present whenever the complexity-specification criterion attributes design. The conclusion of this argument is a straightforward inductive generalization. It has the same logical status as concluding that all ravens are black given that all ravens observed to date have been found to be black.

Anyone with a prior commitment to naturalism is likely to object at this point, claiming that the only things we can know to be designed are artifacts manufactured by intelligent beings that are in turn the product of blind evolutionary processes (for instance, humans). Hence to use the complexity-specification criterion to extrapolate design beyond such artifacts is illegitimate. This argument does not work. It is circular reasoning to invoke naturalism to underwrite an evolutionary account of intelligence and then, in turn, to employ this account of intelligence to insulate naturalism from critique. Naturalism is a metaphysical position, not a scientific theory based on evidence. Any account of intelligence it entails is therefore suspect and needs to be subjected to independent checks. The complexity-specification criterion provides one such check.

If we set aside the naturalist's evolutionary account of intelligence, a more serious objection remains. I am arguing inductively that the complexity-specification criterion is a reliable criterion for detecting design. The conclusion of this argument is that whenever the criterion attributes design, design actually is present. The premise of this argument is that whenever the criterion attributes design and the underlying causal story can be verified, design actually is present. Now, even though the conclusion follows as an inductive generalization from the premise, the premise itself seems false. There are a lot of coincidences out there that seem best explained without invoking design. Consider, for instance, the Shoemaker-Levy comet. The Shoemaker-Levy comet crashed into Jupiter exactly twenty-five years to the day after the Apollo 11 moon landing. What are we to make of this coincidence? Do we really want to explain it in terms of design? What if we submitted this coincidence to the complexity-specification criterion and out popped design? Our intuitions strongly suggest that the comet's trajectory and NASA's space program were operating independently and that at best this coincidence should be referred to chance—certainly not design.

This objection is readily met. The fact is that the complexity-specification criterion does not yield design all that easily, especially if the complexities are kept high (or, correspondingly, the probabilities are kept small). It is simply not the case that unusual and striking coincidences automatically yield design.

Martin Gardner is no doubt correct when he notes, "The number of events in which you participate for a month, or even a week, is so huge that the probability of noticing a startling correlation is quite high, especially if you keep a sharp outlook."[19] The implication he means to draw, however, is incorrect, namely, that therefore startling correlations/coincidences may uniformly be relegated to chance. Yes, the fact that the Shoemaker-Levy comet crashed into Jupiter exactly twenty-five years to the day after the Apollo 11 moon landing is a coincidence best referred to chance. But the fact that Mary Baker Eddy's writings on Christian Science bear a remarkable resemblance to Phineas Parkhurst Quimby's writings on mental healing is a coincidence that cannot be explained by chance and is properly explained by positing Quimby as a source for Eddy.[20]

The complexity-specification criterion is robust and easily resists counterexamples of the Shoemaker-Levy variety. Assuming, for instance, that the Apollo 11 moon landing serves as a specification for the crash of Shoemaker-Levy into Jupiter (a generous concession at that), and that the comet could have crashed at any time within a period of a year, and that the comet crashed to the very second precisely twenty-five years after the moon landing, a straightforward probability calculation indicates that the probability of this coincidence is no smaller than 1 in 10^8. This simply is not all that small a probability (that is, high complexity), especially when considered in relation to all the events astronomers are observing in the solar system. Certainly this probability is nowhere near the universal probability bound of 1 in 10^{150} that I propose in *The Design Inference*.[21] I have yet to see a convincing application of the complexity-specification criterion in which coincidences better explained by chance get attributed to design.

There is one last potential counterexample we need to consider, and that is the possibility of an evolutionary algorithm producing specified complexity. By an evolutionary algorithm I mean any clearly defined procedure that generates contingency via some chance process and then sifts the so-generated contingency via some law-like (that is, necessitarian) process. The Darwinian mutation-selection mechanism, neural nets, and genetic algorithms all fall within this definition of evolutionary algorithms.

Now, it is widely held that evolutionary algorithms are just the means for generating specified complexity apart from design. Yet this widely held view is incorrect. The problem is that evolutionary algorithms cannot generate complexity. This may seem counterintuitive, but consider a well-known example by Richard Dawkins in which he purports to show how a cumulative selection process acting on chance can generate specified complexity.[22] He starts with the target sequence

METHINKS·IT·IS·LIKE·A·WEASEL

(he considers only capital Roman letters and spaces, here represented by bullets—thus 27 possibilities at each location in a symbol string).

If we tried to attain this target sequence by pure chance (for example, by randomly shaking out Scrabble pieces), the probability of getting it on the first try would be around 1 in 10^{40}, and correspondingly it would take on average about 10^{40} tries to stand a better than even chance of getting it. Thus, if we depended on pure chance to attain this target sequence, we would in all likelihood be unsuccessful (granted, this 1 in 10^{40} improbability falls short of my universal probability bound of 1 in 10^{150}, but for practical purposes 1 in 10^{40} is small enough to preclude chance and, yes, implicate design). As a problem for pure chance, attaining Dawkins' target sequence is an exercise in generating specified complexity, and it becomes clear that pure chance simply is not up for the task.

But consider next Dawkins' reframing of the problem. In place of pure chance, he considers the following evolutionary algorithm: (1) Start out with a randomly selected sequence of 28 capital Roman letters and spaces, for example,

WDL·MNLT·DTJBKWIRZREZLMQCO·P

(note that the length of Dawkins' target sequence comprises a total of 28 letters and spaces—that is how many letters and spaces there are in METHINKS·IT·IS·LIKE·A·WEASEL); (2) randomly alter all the letters and spaces in this initial randomly generated sequence; (3) whenever an alteration happens to match a corresponding letter in the target sequence, leave it be and randomly alter only those remaining letters that still differ from the target sequence.

In very short order this algorithm converges to Dawkins' target sequence. In his book *The Blind Watchmaker*, Dawkins provides the following computer simulation of this algorithm:[23]

(1) WDL·MNLT·DTJBKWIRZREZLMQCO·P

(2) WDLTMNLT·DTJBSWIRZREZLMQCO·P

. . .

(10) MDLDMNLS·ITJISWHRZREZ·MECS·P

. . .

(20) MELDINLS·IT·ISWPRKE·Z·WECSEL

. . .

(30) METHINGS·IT·ISWLIKE·B·WECSEL

. . .

(40) METHINKS·IT·IS·LIKE·I·WEASEL

. . .

(43) METHINKS·IT·IS·LIKE·A·WEASEL

Thus, in place of 10^{40} tries on average for pure chance to generate Dawkins' target sequence, it now takes only on average 40 tries to generate it via an evolutionary algorithm. Although Dawkins has gotten a lot of mileage out of this example, exactly what it establishes is very different from what he and much of the evolutionary community think it establishes.[24] For one thing, choosing a target sequence is a deeply teleological move (the target is set prior to running the evolutionary algorithm, and the evolutionary algorithm is explicitly programmed to end up in the target). This is a problem because evolutionary algorithms are supposed to be devoid of teleology. But let us for the sake of argument bracket this teleological problem, which in the case of Darwinism amounts to nature having to select its own targets.

A more serious problem then remains. We can see it by posing the following question: Given Dawkins' evolutionary algorithm, what besides the target sequence can this algorithm attain? Think of it this way. Dawkins' evolutionary algorithm is chugging along; what are the possible terminal points of this algorithm? Clearly, the algorithm is always going to converge on the target sequence (with probability 1 for that matter!). An evolutionary algorithm acts as a *probability amplifier*. Whereas it would take pure chance on average 10^{40} tries to attain Dawkins' target sequence, his evolutionary algorithm on average gets it for you in the logarithm of the number of tries that it takes pure chance, that is, on average in only 40 tries (and with virtual certainty in a few hundred tries).

But a probability amplifier is also a *complexity attenuator*. Recall that the "complexity" in the complexity-specification criterion coincides with improbability. Dawkins' evolutionary algorithm vastly increases the probability of getting the target sequence but in so doing vastly decreases the complexity inherent in the target sequence. The target sequence, if it had to be obtained by randomly throwing Scrabble pieces, would be highly improbable and on average would require a vast number of iterations before it could be obtained. But with Dawkins' evolutionary algorithm, the probability of obtaining the target sequence is high given only a few iterations. In effect, Dawkins' evolutionary algorithm skews the probabilities so that what at first blush seems highly improbable or complex is nothing of the sort. It follows that evolutionary algorithms cannot generate true complexity but only the *appearance of complexity*. And since they cannot generate complexity, they cannot generate specified complexity either.

6. Why the Criterion Works

My second argument for showing that specified complexity reliably detects design considers the nature of intelligent agency and, specifically, what it is about intelligent agents that makes them detectable. Even though induction confirms that specified complexity is a reliable criterion for detecting design, induction does not explain why this criterion works. To see why the complexity-specification criterion is exactly the right instrument for detecting design, we need to understand what it is about intelligent agents that makes them detectable in the first place. The principal characteristic of intelligent agency is *choice*. Even the etymology of the word "intelligent" makes this clear. "Intelligent" derives from two Latin words, the preposition *inter*, meaning between, and the verb *lego*, meaning to choose or select. Thus, according to its etymology, intelligence consists in *choosing between*. For an intelligent agent to act is therefore to choose from a range of competing possibilities.

This is true not just of humans but of animals as well as of extraterrestrial intelligences. A rat navigating a maze must choose whether to go right or left at various points in the maze. When SETI researchers attempt to discover intelligence in the extraterrestrial radio transmissions they are monitoring, they assume an extraterrestrial intelligence could have chosen any number of possible radio transmissions, and then they attempt to match the transmissions they observe with certain patterns as opposed to others. Whenever a human being utters meaningful speech, a choice is made from a range of possible sound combinations that might have been uttered. Intelligent agency always entails discrimination, choosing certain things, ruling out others.

Given this characterization of intelligent agency, the crucial question is how to recognize it. Intelligent agents act by making a choice. How, then, do we recognize that an intelligent agent has made a choice? A bottle of ink spills accidentally onto a sheet of paper; someone takes a fountain pen and writes a message on a sheet of paper. In both instances ink is applied to paper. In both instances one among an almost infinite set of possibilities is realized. In both instances a contingency is actualized and others are ruled out. Yet in one instance we ascribe agency, in the other chance.

What is the relevant difference? Not only do we need to observe that a contingency was actualized, but we ourselves need also to be able to specify that contingency. Alternatively, we need to observe the occurrence of an event that happened but did not have to happen (that is, a contingency), and we must show that this event conforms to a pattern that could be constructed independently of the event (that is, a specification). Ascribing intelligent agency therefore requires identifying both contingency and specification. A random ink blot is contingent but unspecified; a message written with ink on paper is both contingent and specified. To be sure, the exact message recorded may not be specified. But orthographic, syntactic, and semantic constraints will nonetheless specify it.

Actualizing one among several competing possibilities, ruling out the rest, and specifying the one that was actualized encapsulates how we recognize intelligent agency, or, equivalently, how we detect design. Experimental psychologists who study animal learning and behavior have known this all along. To learn a task, an animal must acquire the ability to actualize behaviors suitable for the task as well as the ability to rule out behaviors unsuitable for the task. Moreover, for a psychologist to recognize that an animal has learned a task, it is necessary not only to observe the animal making the appropriate discrimination but also to specify the discrimination.

Thus, to recognize whether a rat has successfully learned how to traverse a maze, a psychologist must first specify which sequence of right and left turns conducts the rat out of the maze. No doubt, a rat randomly wandering a maze also discriminates a sequence of right and left turns. But by randomly wandering the maze, the rat gives no indication that it can discriminate the appropriate sequence of right and left turns for exiting the maze. Consequently, the psychologist studying the rat will have no reason to think the rat has learned how to traverse the maze.

Only if the rat executes the sequence of right and left turns specified by the psychologist will the psychologist recognize that the rat has learned how to traverse the maze. Now it is precisely the learned behaviors we regard as intelligent in animals. Hence it is no surprise that the same scheme for recognizing animal learning recurs for recognizing intelligent agency generally, to wit: actualizing one among several competing possibilities, ruling out the others, and specifying the one actualized.

Note that complexity is implicit here as well. To see this, consider again a rat traversing a maze, but now take a very simple maze in which two right turns conduct the rat out of the maze. How will a psychologist studying the rat determine whether it has learned to exit the maze? Just putting the rat in the maze will not be enough. Because the maze is so simple, the rat could by chance just happen to take two right turns and thereby exit the maze. The psychologist will therefore be uncertain whether the rat actually learned to exit this maze or whether the rat just got lucky.

But contrast this with a complicated maze in which a rat must take just the right sequence of left and right turns to exit the maze. Suppose the rat must take one hundred appropriate right and left turns and that any mistake will prevent the rat from exiting the maze. A psychologist who sees the rat take no erroneous turns and in short order exit the maze will be convinced that the rat has indeed learned how to exit the maze and that this was not dumb luck.

This general scheme for recognizing intelligent agency is but a thinly disguised form of the complexity-specification criterion. In general, to recognize intelligent agency we must observe an actualization of one among several competing possibilities, note which possibilities were ruled out, and then be able to specify the possibility that was actualized. What is more, the competing possibilities that were ruled out must be live possibilities and sufficiently numerous so that specifying the possibility that was actualized cannot be attributed to chance. In terms of complexity, this is just another way of saying that the range of possibilities is complex. In terms of probability, this is just another way of saying that the possibility that was actualized has small probability.

All the elements in this general scheme for recognizing intelligent agency (that is, actualizing, ruling out, and specifying) find their counterpart in the complexity-specification criterion. It follows that this criterion makes precise what we have been doing right along when we recognize intelligent agency. The complexity-specification criterion pinpoints how we detect design.

7. Conclusion

Albert Einstein once said that in science things should be made as simple as possible but no simpler. The materialistic philosophy of science that dominated the end of the nineteenth and much of the twentieth century insists that all phenomena can be explained simply by reference to chance and/or necessity. Nevertheless, this essay has suggested, in effect, that materialistic philosophy portrays reality too simply. There are some entities and events that we cannot and, indeed, do not explain by reference to these twin modes of materialistic causation. Specifically, I have shown that when we encounter entities or events that manifest the joint properties of complexity and specification we routinely, and properly, attribute them, not to chance and/or physical/chemical necessity, but to intelligent design, that is, to mind rather than matter. Clearly, we find the complexity-specification criteria in objects that other human minds have designed. Nevertheless, this essay has not sought to answer the question of whether the criteria that reliably indicate the activity of a prior intelligent mind exist in the natural world, that is, in things that we know humans did not design, such as living organisms or the fundamental architecture of the cosmos. In short, I have not addressed the empirical question of whether the natural world, as opposed to the world of human technology, also bears evidence of intelligent design. It is to this question that my colleagues Stephen Meyer and Michael Behe will now turn.

NOTES

[1] Moses Maimonides, *The Guide for the Perplexed*, trans. M. Friedländer (New York: Dover, 1956), p. 188.

[2] Ibid.

[3] Isaac Newton, *Mathematical Principles of Natural Philosophy*, trans. A. Motte, ed. F. Cajori (Berkeley, Calif.: University of California Press, 1978), pp. 543–44.

[4] Pierre Simon de Laplace, *Celestial Mechanics*, 4 vols., trans. N. Bowditch (New York: Chelsea, 1966).

[5] See the introduction to Pierre Simon de Laplace, *A Philosophical Essay on Probabilities*, trans. F. W. Truscott and F. L. Emory (New York: Dover, 1996).

[6] See John S. Bell, *Speakable and Unspeakable in Quantum Mechanics* (Cambridge: Cambridge University Press, 1987).

[7] William A. Dembski, *The Design Inference* (Cambridge: Cambridge University Press, 1998).

[8] See Aristotle, *Metaphysics*, bk. 5, chap. 2, in *The Basic Works of Aristotle*, ed. R. McKeon (New York: Random House, 1941), p. 752.

[9] Francis Bacon, *The Advancement of Learning*, vol. 30 of *Great Books of the Western World*, ed. R. M. Hutchins (Chicago: Encyclopedia Britannica, 1952).

[10] Monod writes, "The cornerstone of the scientific method is the postulate that nature is objective. In other words, the *systematic* denial that 'true' knowledge can be got at by interpreting phenomena in terms of final causes—that is to say, of 'purpose'." In Jacques Monod, *Chance and Necessity* (New York: Vintage, 1972), p. 21.

[11] Jaki writes: "I want no part whatever with the position . . . in which science is surreptitiously taken for a means of elucidating the utterly metaphysical question of purpose." In Stanley Jaki, *Chesterton, A Seer of Science* (Urbana, Ill.: University of Illinois Press, 1986), pp. 139–40, n. 2.

[12] Eliot Marshall, "Medline Searches Turn Up Cases of Suspected Plagiarism", *Science* 279 (January 23, 1998): 473–74.

[13] Charles Darwin, *On the Origin of Species* (1859; reprint, Cambridge: Harvard University Press, 1964), p. 482.

[14] Strictly speaking, in *The Design Inference* I develop a "specification/small probability criterion". This criterion is equivalent to the complexity-specification criterion described here.

[15] Michael Polanyi, "Life Transcending Physics and Chemistry", *Chemical and Engineering News*, August 21, 1967, pp. 54–66; Michael Polanyi, "Life's Irreducible Structure", *Science* 113 (1968): 1308–12; Timothy Lenoir, *The Strategy of Life: Teleology and Mechanics in Nineteenth Century German Biology* (Dordrecht,

Netherlands: Reidel, 1982), pp. 7–8. See also Hubert Yockey, *Information Theory and Molecular Biology* (Cambridge: Cambridge University Press, 1992), p. 335.

[16] Nevertheless, fabrications may themselves derive from the design of those who observe an event after the fact, such as the archer who deliberately draws the pattern after the arrow has landed.

[17] Because specification is so central to inferring design, I need to elaborate on it. For a pattern to count as a specification, the important thing is not when it was identified but whether in a certain well-defined sense it is *independent* of the event it describes. Drawing a target around an arrow already embedded in a wall is not independent of the arrow's trajectory. Consequently, such a pattern cannot be used to attribute the arrow's trajectory to design. Patterns that are specifications cannot simply be read off the events whose design is in question—in other words, it is not enough to identify a pattern simply by inspecting an event and noting (i.e., "reading off") its features. Rather, to count as specifications, patterns must be suitably independent of events. I refer to this relation of independence as *detachability* and say that a pattern is detachable if and only if it satisfies that relation.

Detachability can be understood as asking the following question: Given an event whose design is in question and a pattern describing it, would we be able to construct that pattern if we had no knowledge which event occurred? Here is the idea. An event has occurred. A pattern describing the event is given. The event is one from a range of possible events. If all we knew was the range of possible events without any specifics about which event actually occurred (e.g., we know that tomorrow's weather will be rain or shine, but we do not know which), could we still construct the pattern describing the event? If so, the pattern is detachable from the event.

For a better picture of what is at stake, I want to consider the following example. It was this example that finally clarified for me what transforms a pattern *simpliciter* into a pattern *qua* specification. Consider, therefore, the following event E, an event that to all appearances was obtained by flipping a fair coin 100 times:

```
THTTTHHTHHTTTTTTTITITIITTIHHHTT
HTHHHTHHHTTTTTTTTHTTHTTTHH
THTTTHTHTHTHHTTHHHHHTTTHTTHH
THTHTHHHHHTTHHTHHHHTHHHHTT          E
```

Is E the product of chance or not? A standard trick of statistics professors with an introductory statistics class is to divide the class in two, having students in one-half of the class each flip a coin 100 times, writing down the sequence of heads and tails on a slip of paper, and having students in the other half each generate purely with their minds a "random-looking" string of coin tosses that mimics the tossing of a coin 100 times, also writing down the sequence of heads and tails on a slip of paper. When the students then hand in their slips of paper, it is the professor's job to sort the papers into two piles, those generated by flipping a fair coin and those concocted in the students' heads. To the amazement of the students, the statistics professor is typically able to sort the papers with 100 percent accuracy.

There is no mystery here. The statistics professor simply looks for a repetition of six or seven heads or tails in a row to distinguish the truly random from the pseudo-random sequences. In 100 coin flips, one is quite likely to see six or seven such repetitions. On the other hand, people concocting pseudo-random sequences with their minds tend to alternate between heads and tails too frequently. Whereas with a truly random sequence of coin tosses there is a 50 percent chance that one toss will differ from the next, as a matter of human psychology people expect that one toss will differ from the next around 70 percent of the time.

How, then, will our statistics professor fare when confronted with the event E described above? Will E be attributed to chance or to the musings of someone trying to mimic chance? According to the professor's crude randomness checker, E would be assigned to the pile of sequences presumed to be truly random, for E contains a repetition of seven tails in a row. Everything that at first blush would lead us to regard E as truly random checks out. There are exactly 50 alternations between heads and tails (as opposed to the 70 that would be expected from humans trying to mimic chance). What's more, the relative frequencies of heads and tails check out: there were 49 heads and 51 tails. Thus it is not as though the coin supposedly responsible for generating E was heavily biased in favor of one side versus the other.

Suppose, however, that our statistics professor suspects she is not up against a neophyte statistics student but instead a fellow statistician who is trying to put one over on her. To help organize her problem, study it more carefully, and enter it into a computer, she will find it convenient to let strings of 0's and 1's represent the outcomes of coin flips, with 1 corresponding to heads and 0 to tails. In that case the following pattern D will correspond to the event E:

0100011011000000101001100

1011101110000000100100011

0100010101100111100010011

0101011110011011110111100   D

Now, the mere fact that the event E conforms to the pattern D is no reason to think that E did not occur by chance. As things stand, the pattern D has simply been read off the event E.

But D need not have been read off of E. Indeed, D could have been constructed without recourse to E. To see this, let us rewrite D as follows:

0

1

00

01

10
11
000
001
010
011
100
101
110
111
0000
0001
0010
0011
0100
0101
0110
0111
1000
1001
1010
1011
1100
1101
1110
1111
00   D

By viewing D this way, anyone with the least exposure to binary arithmetic immediately recognizes that D was constructed simply by writing binary numbers in ascending order, starting with the one-digit binary numbers (i.e., 0 and 1), proceeding then to the two-digit binary numbers (i.e., 00, 01, 10, and 11), and continuing on until 100 digits were recorded. It is therefore intuitively clear that D does not describe a truly random event (i.e., an event obtained by tossing a fair coin) but rather a pseudo-random event, concocted by doing a little binary arithmetic.

Although it is now intuitively clear why chance cannot properly explain E, we need to consider more closely why this is so. We started with a putative chance event E, supposedly obtained by flipping a fair coin 100 times. Since heads and tails each have probability 1/2, and since this probability gets multiplied for each flip of the coin, it follows that the probability of E is 1 in 2^{100}, or approximately 1 in 10^{30} (i.e., one in a thousand billion billion billion). In addition, we constructed a pattern D to which E conforms. Initially D proved insufficient to eliminate chance as the explanation of E since in its construction D was simply read off of E. Rather, to eliminate chance we had also to recognize that D exhibited a pattern independent of E (independence in this case meaning that D could have been constructed quite easily by performing some simple arithmetic operations with binary numbers). Thus, to eliminate chance we needed to employ additional *side information*, which in this case consisted of our knowledge of binary arithmetic. This side information enabled us to establish that D is independent of E (cf. the archer analogy, where the pattern functions as a fixed target that is independent of the arrow's trajectory). Consequently, this side information detaches the pattern D from the event E and thereby renders D a specification.

For side information to detach a pattern from an event, it must satisfy two conditions, a *conditional independence condition* and a *tractability condition*. According to the conditional independence condition, the side information must be conditionally independent of the event E. Conditional independence is a well-defined notion from probability theory. It means that the probability of E does not change once the side information is taken into account. Conditional independence is the standard probabilistic way of unpacking epistemic independence. Two things are epistemically independent if knowledge about one thing (in this case the side information) does not affect knowledge about the other (in this case the occurrence of E). This is certainly the case here, since our knowledge of binary arithmetic does not affect the probabilities of coin tosses.

The second condition, the tractability condition, requires that the side information enable us to construct the pattern D to which E conforms. This is evidently the case here as well, since our knowledge of binary arithmetic enables us to arrange binary numbers in ascending order and thereby construct the pattern D. But what exactly is this *ability to construct a pattern on the basis of side information*? Perhaps the most slippery words in philosophy are "can", "able", and "enable". Fortunately, just as there is a precise theory for characterizing the epistemic independence between an event and side information—namely, probability theory—so too there is a precise theory for characterizing the ability to construct a pattern on the basis of side information—namely, complexity theory.

Complexity theory, conceived now quite generally and not merely as a form of probability, assesses the difficulty of tasks given the resources available for accomplishing those tasks [see chap. 4 of my *The Design Inference*]. As a generalization of computational complexity theory, complexity theory ranks tasks according to difficulty and then determines which tasks are sufficiently manageable to be doable or tractable. For instance, given current technology we find sending a person to the moon tractable but sending a person to the nearest galaxy intractable. In the tractability condition, the task to be accomplished is the construction of a pattern and the resource for accomplishing that task is side information. Thus, for the tractability condition to be satisfied, side information must provide the resources necessary for constructing the pattern in question. All of this admits a precise complexity-theoretic formulation and makes definite what I called "the ability to construct a pattern on the basis of side information".

Taken jointly, the tractability and conditional independence conditions mean that side information enables us to construct the pattern to which an event conforms, yet without recourse to the actual event. This is the crucial insight. Because the side information is conditionally and therefore epistemically independent of the event, any pattern constructed from this side information is obtained without recourse to the event. In this way any pattern that is constructed from such side information avoids the charge of being ad hoc. These, then, are the detachable patterns. These are the specifications.

[18] Dembski, *Design Inference*, chap. 5.

[19] Martin Gardner, "Arthur Koestler: Neoplatonism Rides Again", *World*, August 1, 1972, pp. 87–89.

[20] Walter Martin, *The Kingdom of the Cults*, rev. ed. (Minneapolis. Bethany House, 1985), pp. 127–30.

[21] Dembski, *Design Inference*, chap. 6, sec. 5.

[22] Richard Dawkins, *The Blind Watchmaker* (New York: Norton, 1986), pp. 47–48.

[23] Ibid.

[24] Cf. Bernd-Olaf Küppers, "On the Prior Probability of the Existence of Life", in *The Probabilistic Revolution*, vol. 2, ed. L. Krüger, G. Gigerenzer, and M. S. Morgan (Cambridge: MIT Press, 1987), pp. 355–69. Küppers agrees that Dawkins' METHINKS·IT·IS·LIKE·A·WEASEL example grasps an essential feature of the Darwinian mechanism.

STEPHEN C. MEYER

EVIDENCE FOR DESIGN
IN PHYSICS AND BIOLOGY:
FROM THE ORIGIN OF THE UNIVERSE
TO THE ORIGIN OF LIFE

1. Introduction

In the preceding essay, mathematician and probability theorist William Dembski notes that human beings often detect
the prior activity of rational agents in the effects they leave behind.[1] Archaeologists assume, for example, that rational agents
produced the inscriptions on the Rosetta Stone; insurance fraud
investigators detect certain "cheating patterns" that suggest in-
tentional manipulation of circumstances rather than "natural"
disasters; and cryptographers distinguish between random signals and those that carry encoded messages.

More importantly, Dembski's work establishes the criteria by
which we can recognize the effects of rational agents and distinguish them from the effects of natural causes. In brief, he shows
that systems or sequences that are both "highly complex" (or
very improbable) and "specified" are always produced by intelligent agents rather than by chance and/or physical-chemical
laws. Complex sequences exhibit an irregular and improbable arrangement that defies expression by a simple formula or
algorithm. A specification, on the other hand, is a match or
correspondence between an event or object and an independently given pattern or set of functional requirements.

As an illustration of the concepts of complexity and specification, consider the following three sets of symbols:

⌈   "inetehnsdysk]idmhcpew,ms.s/a"
           "Time and tide wait for no man."
⌊   "ABABABABABABABABABABAB"

Both the first and second sequences shown above are complex because both defy reduction to a simple rule. Each represents a highly irregular, aperiodic, and improbable sequence of symbols. The third sequence is not complex but is instead highly ordered and repetitive. Of the two complex sequences, only the second exemplifies a set of independent functional requirements—that is, only the second sequence is specified. English has a number of functional requirements. For example, to convey meaning in English one must employ existing conventions of vocabulary (associations of symbol sequences with particular objects, concepts, or ideas), syntax, and grammar (such as "every sentence requires a subject and a verb"). When arrangements of symbols "match" or utilize existing vocabulary and grammatical conventions (that is, functional requirements) communication can occur. Such arrangements exhibit "specification". The second sequence ("Time and tide wait for no man") clearly exhibits such a match between itself and the preexisting requirements of English vocabulary and grammar.

Thus, of the three sequences above only the second manifests complexity and specification, both of which must be present for us to infer a designed system according to Dembski's theory. The third sequence lacks complexity, though it does exhibit a simple pattern, a specification of sorts. The first sequence is complex but not specified, as we have seen. Only the second sequence, therefore, exhibits both complexity and specification. Thus, according to Dembski's theory, only the second sequence indicates an intelligent cause—as indeed our intuition tells us.

As the above illustration suggests, Dembski's criteria of specification and complexity bear a close relationship to certain concepts of information. As it turns out, the joint criteria of complexity and specification (or "specified complexity") are equivalent or "isomorphic" with the term "information content",[2] as it is often used.[3] Thus, Dembski's work suggests that "high information content" indicates the activity of an intelligent agent. Common, as well as scientific, experience confirms this theoretical insight. For example, few rational people would attribute hieroglyphic inscriptions to natural forces such as wind or erosion rather than to intelligent activity.

Dembski's work also shows how we use a comparative reasoning process to decide between natural and intelligent causes. We usually seek to explain events by reference to one of three competing types of explanation: chance, necessity (as the result of physical-chemical laws), and/or design, (that is, as the work of an intelligent agent). Dembski has created a formal model of evaluation that he calls "the explanatory filter". The filter shows that the best explanation of an event is determined by its probabilistic features or "signature". Chance best explains events of small or intermediate probability; necessity (or physical-chemical law) best explains events of high probability; and intelligent design best explains small probability events that also manifest specificity (of function, for example). His "explanatory filter" constitutes, in effect, a scientific method for detecting the activity of intelligence. When events are both highly improbable and specified (by an independent pattern) we can reliably detect the activity of intelligent agents. In such cases, explanations involving design are better than those that rely exclusively on chance and/or deterministic natural processes.

Dembski's work shows that detecting the activity of intelligent agency ("inferring design") represents an indisputably common form of rational activity. His work also suggests that the properties of complexity and specification reliably indicate the prior activity of an intelligent cause. This essay will build on this insight to address another question. It will ask: Are the criteria that indicate intelligent design present in features of nature that clearly preexist the advent of humans on earth? Are the features that indicate the activity of a designing intel-

56 STEPHEN C. MEYER

ligence present in the physical structure of the universe or in the features of living organisms? If so, does intelligent design still constitute the best explanation of these features, or might naturalistic explanations based upon chance and/or physico-chemical necessity constitute a better explanation? This paper will evaluate the merits of the design argument in light of developments in physics and biology as well as Dembski's work on "the design inference". I will employ Dembski's comparative explanatory method (the "explanatory filter") to evaluate the competing explanatory power of chance, necessity, and design with respect to evidence in physics and biology. I will argue that intelligent design (rather than chance, necessity, or a combination of the two) constitutes the best explanation of these phenomena. I will, thus, suggest an empirical, as well as a theoretical, basis for resuscitating the design argument.

2.1 Evidence of Design in Physics:
Anthropic "Fine Tuning"

Despite the long popularity of the design argument in the history of Western thought, most scientists and philosophers had come to reject the design argument by the beginning of the twentieth century. Developments in philosophy during the eighteenth century and developments in science during the nineteenth (such as Laplace's nebular hypothesis and Darwin's theory of evolution by natural selection) left most scientists and scholars convinced that nature did not manifest unequivocal evidence of intelligent design.

During the last forty years, however, developments in physics and cosmology have placed the word "design" back in the scientific vocabulary. Beginning in the 1960s, physicists unveiled a universe apparently fine-tuned for the possibility of human life. They discovered that the existence of life in the universe depends upon a highly improbable but precise balance of physical factors.[4] The constants of physics, the initial conditions of the universe, and many other of its features appear delicately balanced to allow for the possibility of life. Even very slight alterations in the values of many factors, such as the expansion rate of the universe, the strength of gravitational or electromagnetic attraction, or the value of Planck's constant, would render life impossible. Physicists now refer to these factors as "anthropic coincidences" (because they make life possible for man) and to the fortunate convergence of all these coincidences as the "fine tuning of the universe". Given the improbability of the precise ensemble of values represented by these constants, and their specificity relative to the requirements of a life-sustaining universe, many physicists have noted that the fine tuning strongly suggests design by a preexistent intelligence. As well-known British physicist Paul Davies has put it, "the impression of design is overwhelming."[5]

To see why, consider the following illustration. Imagine that you are a cosmic explorer who has just stumbled into the control room of the whole universe. There you discover an elaborate "universe-creating machine", with rows and rows of dials, each with many possible settings. As you investigate, you learn that each dial represents some particular parameter that has to be calibrated with a precise value in order to create a universe in which life can exist. One dial represents the possible settings for the strong nuclear force, one for the gravitational constant, one for Planck's constant, one for the ratio of the neutron mass to the proton mass, one for the strength of electromagnetic attraction, and so on. As you, the cosmic explorer, examine the dials, you find that they could easily have been tuned to different settings. Moreover, you determine by careful calculation that if any of the dial settings were even slightly altered, life would cease to exist. Yet for some reason each dial is set at just the exact value necessary to keep the universe running. What do you infer about the origin of these finely tuned dial settings?

Not surprisingly, physicists have been asking the same question. As astronomer George Greenstein mused, "the thought insistently arises that some supernatural agency, or rather Agency, must be involved. Is it possible that suddenly, without intending to, we have stumbled upon scientific proof for the existence of a Supreme Being? Was it God who stepped in and so providentially crafted the cosmos for our benefit?"[6] For many scientists,[7] the design hypothesis seems the most obvious and intuitively plausible answer to this question. As Sir Fred Hoyle commented, "a commonsense interpretation of the facts suggests that a superintellect has monkeyed with physics, as well as chemistry and biology, and that there are no blind forces worth speaking about in nature."[8] Many physicists now concur. They would argue that, given the improbability and yet the precision of the dial settings, design seems the most plausible explanation for the anthropic fine tuning. Indeed, it is precisely the combination of the improbability (or complexity) of the settings and their specificity relative to the conditions required for a life-sustaining universe that seems to trigger the "commonsense" recognition of design.

2.2 Anthropic Fine Tuning and the Explanatory Filter

Yet several other types of interpretations have been proposed: (1) the so-called weak anthropic principle, which denies that the fine tuning needs explanation; (2) explanations based upon natural law; and (3) explanations based upon chance. Each of these approaches denies that the fine tuning of the universe resulted from an intelligent agent. Using Dembski's "explanatory filter", this section will compare the explanatory power of competing types of explanations for the origin of the anthropic fine tuning. It will also argue, contra (1), that the fine tuning does require explanation.

Of the three options above, perhaps the most popular approach, at least initially, was the "weak anthropic principle" (WAP). Nevertheless, the WAP has recently encountered severe criticism from philosophers of physics and cosmology. Advocates of WAP claimed that if the universe were not fine-tuned to allow for life, then humans would not be here to observe it. Thus, they claimed, the fine tuning requires no explanation. Yet as John Leslie and William Craig have argued, the origin of the fine tuning does require explanation.[9] Though we humans should not be surprised to find ourselves living in a universe suited for life (by definition), we ought to be surprised to learn that the conditions necessary for life are so vastly improbable. Leslie likens our situation to that of a blindfolded man who has discovered that, against all odds, he has survived a firing squad of one hundred expert marksmen.[10] Though his continued existence is certainly consistent with all the marksmen having missed, it does not explain why the marksmen actually did miss. In essence, the weak anthropic principle wrongly asserts that the statement of a necessary condition of an event eliminates the need for a causal explanation of that event. Oxygen is a necessary condition of fire, but saying so does not provide a causal explanation of the San Francisco fire. Similarly, the fine tuning of the physical constants of the universe is a necessary condition for the existence of life, but that does not explain, or eliminate the need to explain, the origin of the fine tuning.

While some scientists have denied that the fine-tuning coincidences require explanation (with the WAP), others have tried to find various naturalistic explanations for them. Of these, appeals to natural law have proven the least popular for a simple reason. The precise "dial settings" of the different constants of physics are specific features *of the laws of nature themselves*. For example, the gravitational constant G determines just how strong gravity will be, given two bodies of known mass separated by a known distance. The constant G is a term *within* the equation that describes gravitational attraction. In this same way, all the constants of the fundamental laws of physics are features of the laws themselves. Therefore, the laws cannot explain these features; they comprise the features that we need to explain.

As Davies has observed, the laws of physics "seem themselves to be the product of exceedingly ingenious design".[11] Further, natural laws by definition describe phenomena that conform to regular or repetitive patterns. Yet the idiosyncratic values of the physical constants and initial conditions of the universe constitute a highly irregular and nonrepetitive ensemble. It seems unlikely, therefore, that any law could explain why all the fundamental constants have exactly the values they do—why, for example, the gravitational constant should have exactly the value 6.67×10^{-11} Newton-meters^2 per kilogram^2 *and* the permittivity constant in Coulombs law the value 8.85×10^{-12} Coulombs^2 per Newton-meter^2, *and* the electron charge to mass ratio 1.76×10^{11} Coulombs per kilogram, *and* Planck's constant 6.63×10^{-34} Joules-seconds, and so on.[12] These values specify a highly complex array. As a group, they do not seem to exhibit a regular pattern that could in principle be subsumed or explained by natural law.

Explaining anthropic coincidences as the product of chance has proven more popular, but this has several severe liabilities as well. First, the immense improbability of the fine tuning makes straightforward appeals to chance untenable. Physicists have discovered more than thirty separate physical or cosmological parameters that require precise calibration in order to produce a life-sustaining universe.[13] Michael Denton, in his book *Nature's Destiny* (1998), has documented many other necessary conditions for specifically human life from chemistry, geology, and biology. Moreover, many individual parameters exhibit an extraordinarily high degree of fine tuning. The expansion rate of the universe must be calibrated to one part in 10^{60}.[14] A slightly more rapid rate of expansion—by one part in 10^{60}—would have resulted in a universe too diffuse in matter to allow stellar formation.[15] An even slightly less rapid rate of expansion—by the same factor—would have produced an immediate gravitational recollapse. The force of gravity itself requires fine tuning to one part in 10^{40}.[16] Thus, our cosmic explorer finds himself confronted not only with a large ensemble of separate dial settings but with very large dials containing a vast array of possible settings, only very few of which allow for a life-sustaining universe. In many cases, the odds of arriving at a single correct setting by chance, let alone all the correct settings, turn out to be virtually infinitesimal. Oxford physicist Roger Penrose has noted that a single parameter, the so-called "original phase-space volume", required such precise fine tuning that the "Creator's aim must have been [precise] to an accuracy of one part in $10^{10^{123}}$" (which is ten billion multiplied by itself 123 times). Penrose goes on to remark that, "one could not possibly even write the number down in full . . . [since] it would be '1' followed by 10^{123} successive 'o's!" —more zeros than the number of elementary particles in the entire universe. Such is, he concludes, "the precision needed to set the universe on its course".[17]

To circumvent such vast improbabilities, some scientists have postulated the existence of a quasi-infinite number of parallel universes. By doing so, they increase the amount of time and number of possible trials available to generate a life-sustaining universe and thus increase the probability of such a universe arising by chance. In these "many worlds" or "possible worlds" scenarios—which were originally developed as part of the "Everett interpretation" of quantum physics and the inflationary Big Bang cosmology of André Linde—any event that could happen, however unlikely it might be, must happen somewhere in some other parallel universe.[18] So long as life has a positive (greater than zero) probability of arising, it had to arise in some possible world. Therefore, sooner or later some universe had to acquire life-sustaining characteristics. Clifford Longley explains that according to the many-worlds hypothesis:

> There could have been millions and millions of different universes created each with different dial settings of the fundamental ratios and constants, so many in fact that the right set was bound to turn up by sheer chance. We just happened to be the lucky ones.[19]

According to the many-worlds hypothesis, our existence in the universe only appears vastly improbable, since calculations about the improbability of the anthropic coincidences arising by chance only consider the "probabilistic resources" (roughly, the amount of time and the number of possible trials) available within our universe and neglect the probabilistic resources available from the parallel universes. According to the many-worlds hypothesis, chance can explain the existence of life in the universe after all.

The many-worlds hypothesis now stands as the most popular naturalistic explanation for the anthropic fine tuning and thus warrants detailed comment. Though clearly ingenious, the many-worlds hypothesis suffers from an overriding difficulty: we have no evidence for any universes other than our own. Moreover, since possible worlds are by definition causally inaccessible to our own world, there can be no evidence for their existence except that they allegedly render probable otherwise vastly improbable events. Of course, no one can observe a designer directly either, although a theistic designer—that is, God —is not causally disconnected from our world. Even so, recent work by philosophers of science such as Richard Swinburne, John Leslie, Bill Craig,[20] Jay Richards,[21] and Robin Collins have established several reasons for preferring the (theistic) design hypothesis to the naturalistic many-worlds hypothesis.

2.3 Theistic Design: A Better Explanation?

First, all current cosmological models involving multiple universes require some kind of mechanism for generating universes. Yet such a "universe generator" would itself require precisely configured physical states, thus begging the question of its initial design. As Collins describes the dilemma:

> In all currently worked out proposals for what this universe generator could be—such as the oscillating big bang and the vacuum

fluctuation models . . .—the "generator" itself is governed by a complex set of laws that allow it to produce universes. It stands to reason, therefore, that if these laws were slightly different the generator probably would not be able to produce any universes that could sustain life.[22]

Indeed, from experience we know that some machines (or factories) can produce other machines. But our experience also suggests that such machine-producing machines themselves require intelligent design.

Second, as Collins argues, all things being equal, we should prefer hypotheses "that are natural extrapolations from what we already know" about the causal powers of various kinds of entities.[23] Yet when it comes to explaining the anthropic coincidences, the multiple-worlds hypothesis fails this test, whereas the theistic-design hypothesis does not. To illustrate, Collins asks his reader to imagine a paleontologist who posits the existence of an electromagnetic "dinosaur-bone-producing field", as opposed to actual dinosaurs, as the explanation for the origin of large fossilized bones. While certainly such a field qualifies as a *possible* explanation for the origin of the fossil bones, we have no experience of such fields or of their *producing* fossilized bones. Yet we have observed animal remains in various phases of decay and preservation in sediments and sedimentary rock. Thus, most scientists rightly prefer the actual dinosaur hypothesis over the apparent dinosaur hypothesis (that is, the "dinosaur-bone-producing-field" hypothesis) as an explanation for the origin of fossils. In the same way, Collins argues, we have no experience of anything like a "universe generator" (that is not itself designed; see above) producing finely tuned systems or infinite and exhaustively random ensembles of possibilities. Yet we do have extensive experience of intelligent agents producing finely tuned machines such as Swiss watches. Thus, Collins concludes, when we postulate "a supermind" (God) to explain the fine tuning of the universe, we are extrapolating from our experience of the causal powers of known entities (that is, intelligent humans), whereas when we postulate the existence of an infinite number of separate universes, we are not.

Third, as Craig has shown, for the many-worlds hypothesis to suffice as an explanation for anthropic fine tuning, it must posit an exhaustively random distribution of physical parameters and thus an *infinite* number of parallel universes to insure that a life-producing combination of factors will eventually arise. Yet neither of the physical models that allow for a multiple-universe interpretation—Everett's quantum-mechanical model or Linde's inflationary cosmology—provides a compelling justification for believing that such an exhaustively random and infinite number of parallel universes exists, but instead only a finite and nonrandom set.[24] The Everett model, for example, only generates an ensemble of material *states*, each of which exists within a parallel universe that has the same set of physical laws and constants as our own. Since the physical constants do not vary "across universes", Everett's model does nothing to increase the probability of the precise fine tuning of constants in our universe arising by chance. Though Linde's model does envision a variable ensemble of physical constants in each of his individual "bubble universes", his model fails to generate either an exhaustively random set of such conditions or the infinite number of universes required to render probable the life-sustaining fine tuning of our universe.

Fourth, Richard Swinburne argues that the theistic-design hypothesis constitutes a simpler and less ad hoc hypothesis than the many-worlds hypothesis.[25] He notes that virtually the only evidence for many worlds is the very anthropic fine tuning the hypothesis was formulated to explain. On the other hand, the theistic-design hypothesis, though also only supported by indirect evidences, can explain many separate and independent features of the universe that the many-worlds scenario cannot, including the origin of the universe itself, the mathematical beauty and elegance of physical laws, and personal religious experience. Swinburne argues that the God hypothesis is a simpler as well as a more comprehensive explanation because it requires the postulation of only one explanatory entity, rather than the multiple entities—including the finely tuned universe generator and the infinite number of causally separate universes —required by the many-worlds hypothesis.

Swinburne and Collins' arguments suggest that few reasonable people would accept such an unparsimonious and far-fetched explanation as the many-worlds hypothesis in any other domain of life. That some scientists dignify the many-worlds hypothesis with serious discussion may speak more to an unimpeachable commitment to naturalistic philosophy than to any compelling merit for the idea itself. As Clifford Longley noted in the *London Times* in 1989,[26] the use of the many-worlds hypothesis to avoid the theistic-design argument often seems to betray a kind of special pleading and metaphysical desperation. As Longley explains:

> The [anthropic-design argument] and what it points to is of such an order of certainty that in any other sphere of science, it would be regarded as settled. To insist otherwise is like insisting that Shakespeare was not written by Shakespeare because it might have been written by a billion monkeys sitting at a billion keyboards typing for a billion years. So it might. But the sight of scientific atheists clutching at such desperate straws has put new spring in the step of theists.[27]

Indeed, it has. As the twentieth century comes to a close, the design argument has reemerged from its premature retirement at the hands of biologists in the nineteenth century. Physics, astronomy, cosmology, and chemistry have each revealed that life depends on a very precise set of design parameters, which, as it happens, have been built into our universe. The fine-tuning evidence has led to a persuasive reformulation of the design hypothesis, even if it does not constitute a formal deductive proof of God's existence. Physicist John Polkinghorne has written that, as a result, "we are living in an age where there is a great revival of natural theology taking place. That revival of natural theology is taking place not on the whole among theologians, who have lost their nerve in that area, but among the scientists."[28] Polkinghorne also notes that this new natural theology generally has more modest ambitions than the natural theology of the Middle Ages. Indeed, scientists arguing for design based upon evidence of anthropic fine tuning tend to do so by inferring an intelligent cause as a "best explanation", rather than by making a formal deductive proof of God's existence. (See Appendix, pp. 213–34, "Fruitful Interchange or Polite Chitchat: The Dialogue between Science and Theology".) Indeed, the foregoing analysis of competing types of causal explanations for the anthropic fine tuning suggests intelligent design precisely as the best explanation for its origin. Thus, fine-tuning evidence may support belief in God's existence, even if it does not "prove" it in a deductively certain way.

3.1 Evidence of Intelligent Design in Biology

Despite the renewed interest in design among physicists and cosmologists, most biologists are still reluctant to consider such notions. Indeed, since the late-nineteenth century, most biologists have rejected the idea that biological organisms manifest evidence of intelligent design. While many acknowledge the appearance of design in biological systems, they insist that purely naturalistic mechanisms such as natural selection acting on random variations can fully account for the appearance of design in living things.

3.2 Molecular Machines

Nevertheless, the interest in design has begun to spread to biology. For example, in 1998 the leading journal, *Cell*, featured a special issue on "Macromolecular Machines". Molecular machines are incredibly complex devices that all cells use to process information, build proteins, and move materials back and forth across their membranes. Bruce Alberts, President of the National Academy of Sciences, introduced this issue with an article entitled, "The Cell as a Collection of Protein Machines". In it, he stated that:

> We have always underestimated cells. . . . The entire cell can be viewed as a factory that contains an elaborate network of interlocking assembly lines, each of which is composed of a set of large protein machines. . . . Why do we call the large protein assemblies that underlie cell function protein *machines*? Precisely because, like machines invented by humans to deal efficiently with the macroscopic world, these protein assemblies contain highly coordinated moving parts.[29]

Alberts notes that molecular machines strongly resemble machines designed by human engineers, although as an orthodox neo-Darwinian he denies any role for actual, as opposed to apparent, design in the origin of these systems.

In recent years, however, a formidable challenge to this view has arisen within biology. In his book *Darwin's Black Box* (1996), Lehigh University biochemist Michael Behe shows that neo-Darwinists have failed to explain the origin of complex molecular machines in living systems. For example, Behe looks at the ion-powered rotary engines that turn the whip-like flagella of certain bacteria.[30] He shows that the intricate machinery in this molecular motor—including a rotor, a stator, O-rings, bushings, and a drive shaft—requires the coordinated interaction of some forty complex protein parts. Yet the absence of any one of these proteins results in the complete loss of motor function. To assert that such an "irreducibly complex" engine emerged gradually in a Darwinian fashion strains credulity. According to Darwinian theory, natural selection selects functionally advantageous systems.[31] Yet motor function only ensues *after* all the necessary parts have independently self-assembled—an astronomically improbable event. Thus, Behe insists that Darwinian mechanisms cannot account for the origin of molecular motors and other "irreducibly complex systems" that require the coordinated interaction of multiple independent protein parts.

To emphasize his point, Behe has conducted a literature search of relevant technical journals.[32] He has found a complete absence of gradualistic Darwinian explanations for the origin of the systems and motors that he discusses. Behe concludes that neo-Darwinists have not explained, or in most cases even attempted to explain, how the appearance of design in "irreducibly complex" systems arose naturalistically. Instead, he notes that we know of only one cause sufficient to produce functionally integrated, irreducibly complex systems, namely, intelligent design. Indeed, whenever we encounter irreducibly complex systems and we know how they arose, they were invariably designed by an intelligent agent. Thus, Behe concludes (on strong uniformitarian grounds) that the molecular machines and complex systems we observe in cells must also have had an intelligent source. In brief, molecular motors appear designed because they were designed.

3.3 The Complex Specificity
of Cellular Components

As Dembski has shown elsewhere,[33] Behe's notion of "irreducible complexity" constitutes a special case of the "complexity" and "specification" criteria that enables us to detect intelligent design. Yet a more direct application of Dembski's criteria to biology can be made by analyzing proteins, the macromolecular components of the molecular machines that Behe examines inside the cell. In addition to building motors and other biological structures, proteins perform the vital biochemical functions —information processing, metabolic regulation, signal transduction—necessary to maintain and create cellular life.

Biologists, from Darwin's time to the late 1930s, assumed that proteins had simple, regular structures explicable by reference to mathematical laws. Beginning in the 1950s, however, biologists made a series of discoveries that caused this simplistic view of proteins to change. Molecular biologist Fred Sanger determined the sequence of constituents in the protein molecule insulin. Sanger's work showed that proteins are made of long nonrepetitive sequences of amino acids, rather like an irregular arrangement of colored beads on a string.[34] Later in the 1950s, work by John Kendrew on the structure of the protein myoglobin showed that proteins also exhibit a surprising three-dimensional complexity. Far from the simple structures that biologists had imagined, Kendrew's work revealed an extraordinarily complex and irregular three-dimensional shape— a twisting, turning, tangled chain of amino acids. As Kendrew explained in 1958, "the big surprise was that it was so irregular . . . the arrangement seems to be almost totally lacking in the kind of regularity one instinctively anticipates, and it is more complicated than has been predicted by any theory of protein structure."[35]

During the 1950s, scientists quickly realized that proteins possess another remarkable property. In addition to their complexity, proteins also exhibit specificity, both as one-dimensional arrays and as three-dimensional structures. Whereas proteins are built from rather simple chemical building blocks known as amino acids, their function—whether as enzymes, signal transducers, or structural components in the cell— depends crucially upon the complex but specific sequencing of these building blocks.[36] Molecular biologists such as Francis Crick quickly likened this feature of proteins to a linguistic text. Just as the meaning (or function) of an English text depends upon the sequential arrangement of letters in a text, so too does the function of a polypeptide (a sequence of amino acids) depend upon its specific sequencing. Moreover, in both cases, slight alterations in sequencing can quickly result in loss of function.

In the biological case, the specific sequencing of amino acids gives rise to specific three-dimensional structures. This structure or shape in turn (largely) determines what function, if any, the amino acid chain can perform within the cell. A functioning protein's three-dimensional shape gives it a "hand-in-glove" fit with other molecules in the cell, enabling it to catalyze specific chemical reactions or to build specific structures within the cell. Due to this specificity, one protein cannot usually substitute for another any more than one tool can substitute for another. A topoisomerase can no more perform the job of a polymerase, than a hatchet can perform the function of a soldering iron. Proteins can perform functions only by virtue of their three-dimensional specificity of fit with other equally specified and complex molecules within the cell. This three-dimensional specificity derives in turn from a one-dimensional specificity of sequencing in the arrangement of the amino acids that form proteins.

3.4. The Sequence Specificity of DNA

The discovery of the complexity and specificity of proteins has raised an important question. How did such complex but specific structures arise in the cell? This question recurred with particular urgency after Sanger revealed his results in the early 1950s. Clearly, proteins were too complex and functionally specific to arise "by chance". Moreover, given their irregularity, it seemed unlikely that a general chemical law or regularity governed their assembly. Instead, as Nobel Prize winner Jacques Monod recalled, molecular biologists began to look for some source of information within the cell that could direct the construction of these highly specific structures. As Monod would later recall, to explain the presence of the specific sequencing of proteins, "you absolutely needed a code."[37]

In 1953, James Watson and Francis Crick elucidated the structure of the DNA molecule.[38] The structure they discovered suggested a means by which information or "specificity" of sequencing might be encoded along the spine of DNA's sugar-phosphate backbone.[39] Their model suggested that variations in sequencing of the nucleotide bases might find expression in the sequencing of the amino acids that form proteins. Francis Crick proposed this idea in 1955, calling it the "sequence hypothesis".[40]

According to Crick's hypothesis, the specific arrangement of the nucleotide bases on the DNA molecule generates the specific arrangement of amino acids in proteins.[41] The sequence hypothesis suggested that the nucleotide bases in DNA functioned like letters in an alphabet or characters in a machine code. Just as alphabetic letters in a written language may perform a communication function depending upon their sequencing, so too, Crick reasoned, the nucleotide bases in DNA may result in the production of a functional protein molecule depending upon their precise sequential arrangement. In both cases, function depends crucially upon sequencing. The nucleotide bases in DNA function in precisely the same way as symbols in a machine code or alphabetic characters in a book. In each case, the arrangement of the characters determines the function of the sequence as a whole. As Dawkins notes, "The machine code of the genes is uncannily computer-like."[42] Or, as software innovator Bill Gates explains, "DNA is like a computer program, but far, far more advanced than any software we've ever created."[43] In the case of a computer code, the specific arrangement of just two symbols (0 and 1) suffices to carry information. In the case of an English text, the twenty-six letters of the alphabet do the job. In the case of DNA, the complex but precise sequencing of the four nucleotide bases adenine, thymine, guanine, and cytosine (A, T, G, and C) —stores and transmits genetic information, information that finds expression in the construction of specific proteins. Thus, the sequence hypothesis implied not only the complexity but also the functional specificity of DNA base sequencing.

4.1 The Origin of Life and the Origin of Biological Information (or Specified Complexity)

Developments in molecular biology have led scientists to ask how the specific sequencing—the information content or specified complexity—in both DNA and proteins originated. These developments have also created severe difficulties for all strictly naturalistic theories of the origin of life. Since the late 1920s, naturalistically minded scientists have sought to explain the origin of the very first life as the result of a completely undirected process of "chemical evolution". In *The Origin of Life* (1938), Alexander I. Oparin, a pioneering chemical evolutionary theorist, envisioned life arising by a slow process of transformation starting from simple chemicals on the early earth. Unlike Darwinism, which sought to explain the origin and diversification of new and more complex living forms from simpler, preexisting forms, chemical evolutionary theory seeks to explain the origin of the very first cellular life. Yet since the late 1950s, naturalistic chemical evolutionary theories have been unable to account for the origin of the complexity and specificity of DNA base sequencing necessary to build a living cell.[44] This section will, using the categories of Dembski's explanatory filter, evaluate the competing types of naturalistic explanations for the origin of specified complexity or information content necessary to the first living cell.

4.2 Beyond the Reach of Chance

Perhaps the most common popular view about the origin of life is that it happened by chance. A few scientists have also voiced support for this view at various times during their careers. In 1954 physicist George Wald, for example, argued for the causal efficacy of chance operating over vast expanses of time. As he stated, "Time is in fact the hero of the plot. . . . Given so much time, the impossible becomes possible, the possible probable, and the probable virtually certain."[45] Later Francis Crick would suggest that the origin of the genetic code—that is, the translation system—might be a "frozen accident".[46] Other theories have invoked chance as an explanation for the origin of genetic information, often in conjunction with prebiotic natural selection. (See section 4.3.)

While some scientists may still invoke "chance" as an explanation, most biologists who specialize in origin-of-life research now reject chance as a possible explanation for the origin of the information in DNA and proteins.[47] Since molecular biologists began to appreciate the sequence specificity of proteins and nucleic acids in the 1950s and 1960s, many calculations have been made to determine the probability of formulating functional proteins and nucleic acids at random. Various methods of calculating probabilities have been offered by Morowitz,[48] Hoyle and Wickramasinghe,[49] Cairns-Smith,[50] Prigogine,[51] and Yockey.[52] For the sake of argument, such calculations have often assumed extremely favorable prebiotic conditions (whether realistic or not), much more time than there was actually available on the early earth, and theoretically maximal reaction rates among constituent monomers (that is, the constituent parts of proteins, DNA and RNA). Such calculations have invariably shown that the probability of obtaining functionally sequenced biomacromolecules at random is, in Prigogine's words, "vanishingly small . . . even on the scale of . . . billions of years".[53] As Cairns-Smith wrote in 1971:

> Blind chance . . . is very limited. Low-levels of cooperation he [blind chance] can produce exceedingly easily (the equivalent of letters and small words), but he becomes very quickly incompetent as the amount of organization increases. Very soon indeed long waiting periods and massive material resources become irrelevant.[54]

Consider the probabilistic hurdles that must be overcome to construct even one short protein molecule of about one hundred amino acids in length. (A typical protein consists of about three hundred amino acid residues, and many crucial proteins are very much longer.)[55]

First, all amino acids must form a chemical bond known as a peptide bond so as to join with other amino acids in the protein chain. Yet in nature many other types of chemical bonds are possible between amino acids; in fact, peptide and nonpeptide bonds occur with roughly equal probability. Thus, at any given site along a growing amino acid chain the probability of having a peptide bond is roughly $1/2$. The probability of attaining four peptide bonds is: $(1/2 \times 1/2 \times 1/2 \times 1/2) = 1/16$ or $(1/2)^4$. The probability of building a chain of one hundred amino acids in which all linkages involve peptide linkages is $(1/2)^{99}$, or roughly 1 chance in 10^{30}.

Secondly, in nature every amino acid has a distinct mirror image of itself, one left-handed version, or L-form, and one right-handed version, or D-form. These mirror-image forms are called optical isomers. Functioning proteins use only left-handed amino acids, yet the right-handed and left-handed isomers occur in nature with roughly equal frequency. Taking this into consideration compounds the improbability of attaining a biologically functioning protein. The probability of attaining at random only L-amino acids in a hypothetical peptide chain one hundred amino acids long is $(1/2)^{100}$, or again roughly 1 chance in 10^{30}. The probability of building a one hundred-amino-acid-length chain at random in which all bonds are peptide bonds and all amino acids are L-form would be roughly 1 chance in 10^{60}.

Finally, functioning proteins have a third independent requirement, which is the most important of all; their amino acids must link up in a specific sequential arrangement, just as the letters in a sentence must be arranged in a specific sequence to be meaningful. In some cases, even changing one amino acid at a given site can result in a loss of protein function. Moreover, because there are twenty biologically occurring amino acids, the probability of getting a specific amino acid at a given site is small, that is, 1/20. (Actually the probability is even lower because there are many nonproteineous amino acids in nature.) On the assumption that all sites in a protein chain require one particular amino acid, the probability of attaining a particular protein one hundred amino acids long would be $(1/20)^{100}$, or roughly 1 chance in 10^{130}. We know now, however, that some sites along the chain do tolerate several of the twenty proteineous amino acids, while others do not. The biochemist Robert Sauer of MIT has used a technique known as "cassette mutagenesis" to determine just how much variance among amino acids can be tolerated at any given site in several proteins. His results have shown that, even taking the possibility of variance into account, the probability of achieving a functional sequence of amino acids[56] in several known proteins at random is still "vanishingly small", roughly 1 chance in 10^{65} —an astronomically large number.[57] (There are 10^{65} atoms in our galaxy.)[58]

Moreover, if one also factors in the need for proper bonding and homochirality (the first two factors discussed above), the probability of constructing a rather short functional protein at random becomes so small (1 chance in 10^{125}) as to approach the universal probability bound of 1 chance in 10^{150}, the point at which appeals to chance become absurd given the "probabilistic resources" of the entire universe.[59] Further, making the same calculations for even moderately longer proteins easily pushes these numbers well beyond that limit. For example, the probability of generating a protein of only 150 amino acids in length exceeds (using the same method as above)[60] 1 chance in 10^{180}, well beyond the most conservative estimates for the small probability bound given our multi-billion-year-old universe.[61] In other words, given the complexity of proteins, it is extremely unlikely that a random search through all the possible amino acid sequences could generate even a single relatively short functional protein in the time available since the beginning of the universe (let alone the time available on the early earth). Conversely, to have a reasonable chance of finding a short functional protein in such a random search would require vastly more time than either cosmology or geology allows.

More realistic calculations (taking into account the probable presence of nonproteineous amino acids, the need for vastly longer functional proteins to perform specific functions such as polymerization, and the need for multiple proteins functioning in coordination) only compound these improbabilities —indeed, almost beyond computability. For example, recent theoretical and experimental work on the so-called "minimal complexity" required to sustain the simplest possible living organism suggests a lower bound of some 250 to 400 genes and their corresponding proteins.[62] The nucleotide sequence space corresponding to such a system of proteins exceeds 4^{300000}. The improbability corresponding to this measure of molecular complexity again vastly exceeds 1 chance in 10^{150}, and thus the "probabilistic resources" of the entire universe.[63] Thus, when one considers the full complement of functional biomolecules required to maintain minimal cell function and vitality, one can see why chance-based theories of the origin of life have been abandoned. What Mora said in 1963 still holds:

> Statistical considerations, probability, complexity, etc., followed to their logical implications suggest that the origin and continuance of life is not controlled by such principles. An admission of this is the use of a period of practically infinite time to obtain the derived result. Using such logic, however, we can prove anything.[64]

Though the probability of assembling a functioning biomolecule or cell by chance alone is exceedingly small, origin-of-life researchers have not generally rejected the chance hypothesis merely because of the vast improbabilities associated with these events. Many improbable things occur every day by chance. Any hand of cards or any series of rolled dice will represent a highly improbable occurrence. Yet observers often justifiably attribute such events to chance alone. What justifies the elimination of the chance is not just the occurrence of a highly improbable event, but the occurrence of a very improbable event that also conforms to an independently given or discernible pattern. If someone repeatedly rolls two dice and turns up a sequence such as: 9, 4, 11, 2, 6, 8, 5, 12, 9, 2, 6, 8, 9, 3, 7, 10, 11, 4, 8 and 4, no one will suspect anything but the interplay of random forces, though this sequence does represent a very improbable event given the number of combinatorial possibilities that correspond to a sequence of this length. Yet rolling twenty (or certainly two hundred!) consecutive sevens will justifiably arouse suspicion that something more than chance is in play. Statisticians have long used a method for determining when to eliminate the chance hypothesis that involves prespecifying a pattern or "rejection region".[65] In the dice example above, one could prespecify the repeated occurrence of seven as such a pattern in order to detect the use of loaded dice, for example. Dembski's work discussed above has generalized this method to show how the presence of any conditionally independent pattern, whether temporally prior to the observation of an event or not, can help (in conjunction with a small probability event) to justify rejecting the chance hypothesis.[66]

Origin-of-life researchers have tacitly, and sometimes explicitly, employed precisely this kind of statistical reasoning to justify the elimination of scenarios that rely heavily on chance. Christian de Duve, for example, has recently made this logic explicit in order to explain why chance fails as an explanation for the origin of life:

> A single, freak, highly improbable event can conceivably happen. Many highly improbable events—drawing a winning lottery number or the distribution of playing cards in a hand of bridge—happen all the time. But a string of improbable events—drawing the same lottery number twice, or the same bridge hand twice in a row—does not happen naturally.[67]

De Duve and other origin-of-life researchers have long recognized that the cell represents not only a highly improbable but also a functionally specified system. For this reason, by the mid-1960s most researchers had eliminated chance as a plausible explanation for the origin of the information content or specified complexity necessary to build a cell.[68] Many have instead sought other types of naturalistic explanations (see below).

4.3 Prebiotic Natural Selection: A Contradiction in Terms

Of course, even early theories of chemical evolution did not rely exclusively on chance as a causal mechanism. For example, A. I. Oparin's original theory of evolutionary abiogenesis first published in the 1920s and 1930s invoked prebiotic natural selection as a complement to chance interactions. Oparin's theory envisioned a series of chemical reactions that he thought would enable a complex cell to assemble itself gradually and naturalistically from simple chemical precursors.

For the first stage of chemical evolution, Oparin proposed that simple gases such as ammonia (NH_3), methane (CH_4), water (H_2O), carbon dioxide (CO_2), and hydrogen (H_2) would have rained down to the early oceans and combined with metallic compounds extruded from the core of the earth.[69] With the aid of ultraviolet radiation from the sun, the ensuing reactions would have produced energy-rich hydrocarbon compounds.[70] These in turn would have combined and recombined with various other compounds to make amino acids, sugars, phosphates, and other "building blocks" of the complex molecules (such as proteins) necessary to living cells.[71] These constituents would eventually arrange themselves by chance into primitive metabolic systems within simple cell-like enclosures that Oparin called coacervates.[72] Oparin then proposed a kind of Darwinian competition for survival among his coacervates. Those that, by chance, developed increasingly complex molecules and metabolic processes would have survived to grow more complex and efficient. Those that did not would have dissolved.[73] Thus, Oparin invoked differential survival or natural selection as a mechanism for preserving entities of increasing complexity, thus allegedly helping to overcome the difficulties attendant upon pure-chance hypotheses.

Nevertheless, developments in molecular biology during the 1950s cast doubt on this theory. Oparin originally invoked natural selection to explain how cells refined primitive metabolism once it had arisen. His scenario relied heavily, therefore, on chance to explain the initial formation of the constituent biomacromolecules (such as proteins and DNA) upon which any cellular metabolism would depend. The discovery of the extreme complexity and specificity of these molecules during the 1950s undermined the plausibility of this claim. For this reason, Oparin published a revised version of his theory in 1968 that envisioned a role for natural selection earlier in the process of abiogenesis. His new theory claimed that natural selection acted upon random polymers as they formed and changed within his coacervate protocells.[74] As more complex and efficient molecules accumulated, they would have survived and reproduced more prolifically.

Even so, Oparin's concept of natural selection acting on initially nonliving chemicals (that is, *prebiotic* natural selection) remained problematic. For one thing, it seemed to presuppose a preexisting mechanism of self-replication. Yet self-replication in all extant cells depends upon functional and, therefore, highly sequence-specific proteins and nucleic acids. Yet the origin of specificity in these molecules is precisely what Oparin needed to explain. As Christian de Duve has written, theories of prebiotic natural selection "need information which implies they have to presuppose what is to be explained in the first place".[75] Oparin attempted to circumvent this problem by claiming that the sequences of monomers in the first polymers need not have been highly specific in their arrangement. But this claim raised doubts about whether an accurate mechanism of self-replication (and, thus, natural selection) could have functioned at all. Indeed, Oparin's scenario did not reckon on a phenomenon known as "error catastrophe", in which small "errors" or deviations from functionally necessary sequencing are quickly amplified in successive replications.[76]

Thus, the need to explain the origin of specified complexity in biomacromolecules created an intractable dilemma for Oparin. On the one hand, if he invoked natural selection late in his scenario, then he would in effect attribute the origin of the highly complex and specified biomolecules (necessary to a self-replicating system) to chance alone. Yet, as the mathematician Von Neumann[77] would show, any system capable of self-replication would need to contain subsystems that were functionally equivalent to the information storage, replicating, and processing systems found in extant cells. His calculations and similar ones by Wigner,[78] Landsberg,[79] and Morowitz[80] showed that random fluctuations of molecules in all probability (to understate the case) would not produce the minimal complexity needed for even a primitive replication system.

On the other hand, if Oparin invoked natural selection earlier in the process of chemical evolution, before functional specificity in biomacromolecules had arisen, he could not offer any explanation for how self-replication and thus natural selection could have even functioned. Natural selection presupposes a self-replicating system, but self-replication requires functioning nucleic acids and proteins (or molecules approaching their specificity and complexity)—the very entities Oparin needed to explain. For this reason, the evolutionary biologist Dobzhansky would insist, "prebiological natural selection is a contradiction in terms".[81] Indeed, as a result of this dilemma, most researchers rejected the postulation of prebiotic natural selection as either question begging or indistinguishable from implausible chance-based hypotheses.[82]

Nevertheless, Richard Dawkins[83] and Bernd-Olaf Küppers[84] have recently attempted to resuscitate prebiotic natural selection as an explanation for the origin of biological information. Both accept the futility of naked appeals to chance and invoke what Küppers calls a "Darwinian optimization principle". Both use a computer to demonstrate the efficacy of prebiotic natural selection. Each selects a target sequence to represent a desired functional polymer. After creating a crop of randomly constructed sequences and generating variations among them at random, their computers select those sequences that match the target sequence most closely. The computers then amplify the production of those sequences, eliminate the others (to simulate differential reproduction), and repeat the process. As Küppers puts it, "Every mutant sequence that agrees one bit better with the meaningful or reference sequence . . . will be allowed to reproduce more rapidly."[85] In his case, after a mere thirty-five generations, his computer succeeds in spelling his target sequence, "NATURAL SELECTION".

Despite superficially impressive results, these "simulations" conceal an obvious flaw: molecules in situ do not have a target sequence "in mind". Nor will they confer any selective advantage on a cell, and thus differentially reproduce, until they combine in a functionally advantageous arrangement. Thus, nothing in nature corresponds to the role that the computer plays in selecting functionally nonadvantageous sequences that happen to agree "one bit better" than others with a target sequence. The sequence "NORMAL ELECTION" may agree more with "NATURAL SELECTION" than does the sequence "MISTRESS DEFECTION", but neither of the two yields any advantage in communication over the other, if, that is, we are trying to communicate something about "NATURAL SELECTION". If so, both are equally ineffectual. Similarly, a nonfunctional polypeptide would confer no selective advantage on a hypothetical proto-cell, even if its sequence happens to "agree one bit better" with an unrealized target protein than some other nonfunctional polypeptide.

And, indeed, both Küppers'[86] and Dawkins'[87] published results of their simulations show the early generations of variant phrases awash in nonfunctional gibberish.[88] In Dawkins' simulation, not a single functional English word appears until after the tenth iteration (unlike the more generous example above, which starts with actual, albeit incorrect, words). Yet to make distinctions on the basis of function among sequences that have no function whatsoever would seem quite impossible. Such determination can only be made if considerations of proximity to possible future function are allowed, but this requires foresight that natural selection does not have. But a computer, programmed by a human being, can perform these functions. To imply that molecules can as well only illicitly personifies nature. Thus, if these computer simulations demonstrate anything, they subtly demonstrate the need for intelligent agents to elect some options and exclude others—that is, to create information.

4.4 Self-Organizational Scenarios

Because of the difficulties with chance-based theories, including those that rely upon prebiotic natural selection, most origin-of-life theorists after the mid-1960s attempted to explain the origin of biological information in a completely different way. Researchers began to look for "so-called" self-organizational laws and properties of chemical attraction that might explain the origin of the specified complexity or information content in DNA and proteins. Rather than invoking chance, these theories invoked necessity. Indeed, if neither chance nor prebiotic natural selection acting on chance suffices to explain the origin of large amounts of specified biological information, then scientists committed to finding a naturalistic explanation for the origin of life have needed to rely on principles of physical or chemical necessity. Given a limited number of explanatory options (chance, and/or necessity, or design), the inadequacy of chance has, for many researchers, left only one option. Christian de Duve articulates the logic:

> A string of improbable events—drawing the same lottery number twice, or the same bridge hand twice in a row—does not happen naturally. All of which lead me to conclude that life is an obligatory manifestation of matter, bound to arise where conditions are appropriate.[89]

By the late 1960s origin-of-life biologists began to consider the self-organizational perspective that de Duve describes. At that time, several researchers began to propose that deterministic forces (that is, "necessity") made the origin of life not just probable but inevitable. Some suggested that simple chemicals might possess "self-ordering properties" capable of organizing the constituent parts of proteins, DNA, and RNA into the specific arrangements they now possess.[90] Steinman and Cole, for example, suggested that differential bonding affinities or forces of chemical attraction between certain amino acids might account for the origin of the sequence specificity of proteins.[91] Just as electrostatic forces draw sodium (Na^+) and chloride ions (Cl^-) together into highly ordered patterns within a crystal of salt ($NaCl$), so too might amino acids with special affinities for each other arrange themselves to form proteins. Kenyon and Steinman developed this idea in a book entitled *Biochemical Predestination* in 1969. They argued that life might have been "biochemically predestined" by the properties of attraction that exist between its constituent chemical parts, particularly between the amino acids in proteins.[92]

In 1977, Prigogine and Nicolis proposed another self-organizational theory based on a thermodynamic characterization of living organisms. In *Self-Organization in Nonequilibrium Systems*, Prigogine and Nicolis classified living organisms as open, nonequilibrium systems capable of "dissipating" large quantities of energy and matter into the environment.[93] They observed that open systems driven far from equilibrium often display self-ordering tendencies. For example, gravitational energy will produce highly ordered vortices in a draining bathtub; thermal energy flowing through a heat sink will generate distinctive convection currents or "spiral wave activity". Prigogine and Nicolis argued that the organized structures observed in living systems might have similarly "self-originated" with the aid of an energy source. In essence, they conceded the improbability of simple building blocks arranging themselves into highly ordered structures under normal equilibrium conditions. But

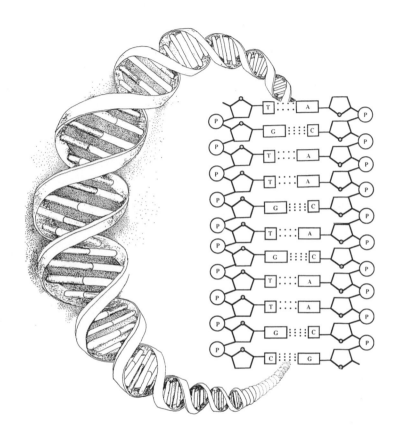

The Bonding Relationships between the Chemical Constituents of the DNA Molecule. Sugars (designated by the pentagons) and phosphates (designated by the circled P's) are linked chemically. Nucleotide bases (A's, T's, G's, and C's) are bonded to the sugar-phosphate backbones. Nucleotide bases are linked by hydrogen bonds (designated by dotted double or triple lines) across the double helix. But no chemical bonds exist between the nucleotide bases along the message-bearing spine of the helix. *Courtesy of Fred Hereen, Daystar Publications.*

they suggested that, under nonequilibrium conditions, where an external source of energy is supplied, biochemical building blocks might arrange themselves into highly ordered patterns.

More recently, Kauffman[94] and de Duve[95] have proposed less detailed self-organizational theories to explain the origin of specified genetic information. Kauffman invokes so-called "autocatalytic properties" that he envisions may emerge from very particular configurations of simple molecules in a rich "chemical minestrone". De Duve envisions proto-metabolism emerging first with genetic information arising later as a by-product of simple metabolic activity. He invokes an extra-evidential principle, his so-called "Cosmic Imperative", to render the emergence of molecular complexity more plausible.

4.5 Order v. Information

For many current origin-of-life scientists self-organizational models now seem to offer the most promising approach to explaining the origin of specified biological information. Nevertheless, critics have called into question both the plausibility and the relevance of self-organizational models. Ironically, a prominent early advocate of self-organization, Dean Kenyon, has now explicitly repudiated such theories as both incompatible with empirical findings and theoretically incoherent [96]

First, empirical studies have shown that some differential affinities do exist between various amino acids (that is, particular amino acids do form linkages more readily with some amino acids than others).[97] Nevertheless, these differences do not correlate to actual sequencing in large classes of known proteins.[98] In short, differing chemical affinities do not explain the multiplicity of amino acid sequences that exist in naturally occurring proteins or the sequential arrangement of amino acids in any particular protein.

In the case of DNA this point can be made more dramatically. The accompanying illustration [p. 84] shows that the structure of DNA depends upon several chemical bonds. There are bonds, for example, between the sugar and the phosphate molecules that form the two twisting backbones of the DNA molecule. There are bonds fixing individual (nucleotide) bases to the sugar-phosphate backbones on each side of the molecule. There are also hydrogen bonds stretching horizontally across the molecule between nucleotide bases making so-called complementary pairs. These bonds, which hold two complementary copies of the DNA message text together, make replication of the genetic instructions possible. Most importantly, however, notice that there are *no* chemical bonds between the bases along the vertical axis in the center of the helix. Yet it is precisely along this axis of the molecule that the genetic information in DNA is stored.[99]

Further, just as magnetic letters can be combined and recombined in any way to form various sequences on a metal surface, so too can each of the four bases A, T, G, and C attach to any site on the DNA backbone with equal facility, making all sequences equally probable (or improbable). Indeed, there are no significant differential affinities between any of the four bases and the binding sites along the sugar-phosphate backbone. The same type of ("n-glycosidic") bond occurs between the base and the backbone regardless of which base attaches. All four bases are acceptable; none is preferred. As Küppers has noted, "the properties of nucleic acids indicate that all the combinatorially possible nucleotide patterns of a DNA are, from a chemical point of view, equivalent."[100] Thus, "self-organizing" bonding affinities cannot explain the sequentially specific arrangement of nucleotide bases in DNA because: (1) there are *no* bonds between bases along the message-bearing axis of the molecule, and (2) there are no *differential* affinities between the backbone and the specific bases that can account for variations in sequencing. Because the same holds for RNA molecules, the theory that life began in an "RNA world" has also failed to solve the sequencing problem[101]—the problem of explaining how specific sequencing in functioning RNA molecules could have arisen in the first place.

For those who want to say that life arose as the result of self-organizing properties intrinsic to the material constituents of living systems, these rather elementary facts of molecular biology have decisive implications. The most obvious place to look for self-organizing properties to explain the origin of genetic information is in the constituent parts of the molecules that carry that information. But biochemistry and molecular biology make clear that forces of attraction between the constituents in DNA, RNA, and proteins do not explain the sequence specificity of these large information-bearing biomolecules.

We know this, in addition to the reasons already stated, because of the multiplicity of variant polypeptides and gene sequences that exist in nature and can be synthesized in the laboratory. The properties of the monomers constituting nucleic acids and proteins simply do not make a particular gene, let alone life as we know it, inevitable. Yet if self-organizational scenarios for the origin of biological information are to have any theoretical import, they must claim just the opposite. And, indeed, they often do, albeit without much specificity. As de Duve has put it, "the processes that generated life" were "highly deterministic", making life as we know it "inevitable" given "the conditions that existed on the prebiotic earth".[102] Yet if we imagine the most favorable prebiotic conditions— a pool of all four DNA nucleotides and all the necessary sugars and phosphates would any particular genetic sequence have to arise? Given all necessary monomers, would any particular functional protein or gene, let alone a specific genetic code, replication system, or signal transduction circuitry, have to arise? Clearly not.

In the parlance of origin-of-life research, monomers are "building blocks". And building blocks can be arranged and rearranged in innumerable ways. The properties of blocks do not determine their arrangement in the construction of buildings. Similarly, the properties of biological *building* blocks do not determine the arrangement of functional polymers. Instead, the chemical properties of the monomers allow for a vast ensemble of possible configurations, the overwhelming majority of which have no biological function whatsoever. Functional genes or proteins are no more inevitable given the properties of their "building blocks" than the palace of Versailles, for example, was inevitable given the properties of the bricks and stone used to construct it. To anthropomorphize, neither bricks and stone nor letters in a written text nor nucleotide bases "care" how they are arranged. In each case, the properties of the constituents remain largely indifferent to the many specific configurations or sequences that they may adopt. Conversely, the properties of nucleotide bases and amino acids do not make any specific sequences "inevitable" as self-organizationalists must claim.

Significantly, information theory makes clear that there is a good reason for this. If chemical affinities between the constituents in the DNA determined the arrangement of the bases, such affinities would dramatically diminish the capacity of DNA to carry information. Consider, for example, what would happen if the individual nucleotide "bases" (A, T, G, C) in a DNA molecule *did* interact by *chemical* necessity with each other. Every time adenine (A) occurred in a growing genetic sequence, it would attract thymine (T) to it.[103] Every time cytosine (C) appeared, guanine (G) would likely follow. As a result, the DNA would be peppered with repetitive sequences of A's followed by T's and C's followed by G's. Rather than a genetic molecule capable of virtually unlimited novelty and characterized by unpredictable and aperiodic sequencing, DNA would contain sequences awash in repetition or redundancy— much like the sequences in crystals. In a crystal the forces of mutual chemical attraction do determine, to a very considerable extent, the sequential arrangement of its constituent parts. As a result, sequencing in crystals is highly ordered and repetitive but neither complex nor informative. Once one has seen

"Na" followed by "Cl" in a crystal of salt, for example, one has seen the extent of the sequencing possible. In DNA, however, where any nucleotide can follow any other, a vast array of novel sequences is possible, corresponding to a multiplicity of amino acid sequences.

The forces of chemical necessity produce redundancy or monotonous order but reduce complexity and thus the capacity to convey novel information. Thus, as the chemist Michael Polanyi noted:

> Suppose that the actual structure of a DNA molecule were due to the fact that the bindings of its bases were much stronger than the bindings would be for any other distribution of bases, then such a DNA molecule would have no information content. Its code-like character would be effaced by an overwhelming redundancy. . . . Whatever may be the origin of a DNA configuration, it can function as a code only if its order is not due to the forces of potential energy. It *must be* as physically indeterminate as the sequence of words is on a printed page [emphasis added].[104]

In other words, if chemists had found that bonding affinities between the nucleotides in DNA produced nucleotide sequencing, they would also have found that they had been mistaken about DNA's information-bearing properties. Or, to put the point quantitatively, to the extent that forces of attraction between constituents in a sequence determine the arrangement of the sequence, to that extent will the information-carrying capacity of the system be diminished or effaced (by redundancy).[105] As Dretske has explained:

> As p(si) [the probability of a condition or state of affairs] approaches 1 the amount of information associated with the occurrence of si goes to 0. In the limiting case when the probability of a condition or state of affairs is unity [p(si) = 1], no information is associated with, or generated by, the occurrence of si. This is merely another way to say that no information is generated by the occurrence of events for which there are no possible alternatives.[106]

Bonding affinities, to the extent they exist, inhibit the max-imization of information[107] because they determine specific outcomes that will follow specific conditions with high prob-ability. Information-carrying capacity is maximized when just the opposite situation obtains, namely, when antecedent con-ditions allow many improbable outcomes.

Of course, the sequences of bases in DNA do not just pos-sess information-carrying capacity or syntactic information or information as measured by classical Shannon information the-ory. These sequences store functionally specified information or specified complexity—that is, they are specified as well as complex. Clearly, however, a sequence cannot be both speci-fied and complex if it is not at least complex. Therefore, the self-organizational forces of chemical necessity that produce redundant order and preclude complexity also preclude the generation of specified complexity (or information content) as well. Chemical affinities do not generate complex sequences. Thus, they cannot be invoked to explain the origin of specified complexity or information content.

The tendency to conflate the qualitative distinctions between "order" and "complexity" has characterized self-organizational research efforts and calls into question the relevance of such work to the origin of life. As Yockey notes, the accumulation of structural or chemical order does not explain the origin of bio-logical complexity or genetic information.[108] He concedes that energy flowing through a system may produce highly ordered patterns. Strong winds form swirling tornadoes and the "eyes" of hurricanes; Prigogine's thermal baths do develop interesting "convection currents"; and chemical elements do coalesce to form crystals. Self-organizational theorists explain well what does not need explaining. What needs explaining in biology is not the origin of order (defined as symmetry or repetition), but the origin of the information content—the highly com-plex, aperiodic, and yet specified sequences that make biolog-ical function possible. As Yockey warns:

Attempts to relate the idea of order . . . with biological organization or specificity must be regarded as a play on words which cannot stand careful scrutiny. Informational macromolecules can code genetic messages and therefore can carry information because the sequence of bases or residues is affected very little, if at all, by [self-organizing] physico-chemical factors.[109]

In the face of these difficulties, some self-organizational theorists have claimed that we must await the discovery of new natural laws to explain the origin of biological information. As Manfred Eigen has argued, "our task is to find an algorithm, a natural law, that leads to the origin of information."[110] But this suggestion betrays confusion on two counts. First, scientific laws do not generally explain or cause natural phenomena, they describe them. For example, Newton's law of gravitation described, but did not explain, the attraction between planetary bodies. Second, laws necessarily describe highly deterministic or predictable relationships between antecedent conditions and consequent events. Laws describe patterns in which the probability of each successive event (given the previous event and the action of the law) becomes inevitable. Yet information mounts as improbabilities multiply. Thus, to say that scientific laws describe complex informational patterns is essentially a contradiction in terms. Instead, scientific laws describe (almost by definition) highly predictable and regular phenomena—that is, redundant order, not complexity (whether specified or otherwise).

5.1 The Return of the Design Hypothesis

If neither chance nor principles of physical-chemical necessity, nor the two acting in combination, explain the ultimate origin of specified complexity or information content in DNA, what does? Do we know of any entity that has the causal powers to create large amounts of specified complexity or information content? We do. As Henry Quastler, an early pioneer in the application of information theory to molecular biology, recognized, the "creation of new information is habitually associated with conscious activity".[111]

Indeed, experience affirms that specified complexity or information content not only routinely arises but always arises from the activity of intelligent minds. When a computer user traces the information on a screen back to its source, he invariably comes to a mind—a software engineer or programmer. If a reader traces the information content in a book or newspaper column back to its source, he will find a writer—again a mental, not a material, cause. Our experientially based knowledge of information confirms that systems with large amounts[112] of specified complexity or information content (especially codes and languages) always originate from an intelligent source— that is, from mental or personal agents. Moreover, this generalization holds not only for the specified complexity or information present in natural languages but also for other forms of specified complexity, whether present in machine codes, machines, or works of art. Like the letters in a section of meaningful text, the parts in a working engine represent a highly improbable and functionally specified configuration. Similarly, the highly improbable shapes in the rock on Mount Rushmore in the United States conform to an independently given pattern—the face of American presidents known from books and paintings. Thus, both these systems have a large amount of *specified* complexity or information content. Not coincidentally, they also resulted from intelligent design, not chance and/or physical-chemical necessity.

This generalization about the cause of specified complexity or information has, ironically, received confirmation from origin-of-life research itself. During the last forty years, every naturalistic model (see n. 44 above) proposed has failed precisely to explain the origin of the specified genetic information required to build a living cell. Thus, mind or intelligence, or what philosophers call "agent causation", now stands as the only cause known to be capable of generating large amounts of specified complexity or information content (from nonbi-ological precursors).

Indeed, because large amounts of specified complexity or information content must be caused by a mind or intelligent design, one can detect the past action of an intelligent cause from the presence of an information-rich effect, even if the cause itself cannot be directly observed.[113] For instance, visitors to the gardens of Victoria harbor in Canada correctly infer the activity of intelligent agents when they see a pattern of red and yellow flowers spelling "Welcome to Victoria", even if they did not see the flowers planted and arranged. Similarly, the specifically arranged nucleotide sequences—the complex but functionally specified sequences—in DNA imply the past action of an intelligent mind, even if such mental agency cannot be directly observed.

Moreover, the logical calculus underlying such inferences follows a valid and well-established method used in all historical and forensic sciences. In historical sciences, knowledge of the present causal powers of various entities and processes enables scientists to make inferences about possible causes in the past. When a thorough study of various possible causes turns up just a single adequate cause for a given effect, historical or forensic scientists can make fairly definitive inferences about the past.[114] Inferences based on knowledge of necessary causes ("distinctive diagnostics") are common in historical and forensic sciences and often lead to the detection of intelligent as well as natural causes. Since criminal X's fingers are the only known cause of criminal X's fingerprints, X's prints on the murder weapon incriminate him with a high degree of certainty. In the same way, since intelligent design is the only known cause of large amounts of specified complexity or information content, the presence of such information implicates an intelligent source.

Scientists in many fields recognize the connection between intelligence and specified complexity and make inferences accordingly. Archaeologists assume a mind produced the inscriptions on the Rosetta Stone. Evolutionary anthropologists argue for the intelligence of early hominids by showing that certain chipped flints are too improbably specified to have been produced by natural causes. NASA's search for extraterrestrial intelligence (SETI) presupposed that the presence of functionally specified information imbedded in electromagnetic signals from space (such as the prime number sequence) would indicate an intelligent source.[115] As yet, however, radio-astronomers have not found such information-bearing signals coming from space. But closer to home, molecular biologists have identified specified complexity or informational sequences and systems in the cell, suggesting, by the same logic, an intelligent cause. Similarly, what physicists refer to as the "anthropic coincidences" constitute precisely a complex and functionally specified array of values. Given the inadequacy of the cosmological explanations based upon chance and law discussed above, and the known sufficiency of intelligent agency as a cause of specified complexity, the anthropic fine-tuning data would also seem best explained by reference to an intelligent cause.

5.2 An Argument from Ignorance?

Of course, many would object that any such arguments from evidence to design constitute arguments from ignorance. Since, these objectors say, we do not yet know how specified complexity in physics and biology could have arisen, we invoke the mysterious notion of intelligent design. On this view, intelligent design functions, not as an explanation, but as a kind of place holder for ignorance.

And yet, we often infer the causal activity of intelligent agents as the best explanation for events and phenomena. As Dembski has shown,[116] we do so rationally, according to clear theoretic criteria. Intelligent agents have unique causal powers that nature does not. When we observe effects that we know from experience only intelligent agents produce, we rightly infer the antecedent presence of a prior intelligence even if we did not observe the action of the particular agent responsible.[117] When these criteria are present, as they are in living systems and in the contingent features of physical law, design constitutes a better explanation than either chance and/or deterministic natural processes.

While admittedly the design inference does not constitute a proof (nothing based upon empirical observation can), it most emphatically does not constitute an argument from ignorance. Instead, the design inference from biological information constitutes an "inference to the best explanation."[118] Recent work on the method of "inference to the best explanation"[119] suggests that we determine which among a set of competing possible explanations constitutes the best one by assessing the causal powers of the competing explanatory entities. Causes that can produce the evidence in question constitute better explanations of that evidence than those that do not. In this essay, I have evaluated and compared the causal efficacy of three broad categories of explanation—chance, necessity (and chance and necessity combined), and design—with respect to their ability to produce large amounts of specified complexity or information content. As we have seen, neither explanations based upon chance nor those based upon necessity, nor (in the biological case) those that combine the two, possess the ability to generate the large amounts of specified complexity or information content required to explain either the origin of life or the origin of the anthropic fine tuning. This result comports with our ordinary and uniform human experience. Brute matter—whether acting randomly or by necessity—does not have the capability to generate novel information content or specified complexity.

Yet it is not correct to say that we do not know how specified complexity or information content arises. We know from experience that conscious intelligent agents can and do create specified information-rich sequences and systems. Furthermore, experience teaches that whenever large amounts of specified complexity or information content are present in an artifact or entity whose causal story is known, invariably creative intelligence—design—has played a causal role in the origin of that entity. Thus, when we encounter such information in the biomacromolecules necessary to life, or in the fine tuning of the laws of physics, we may infer based upon our present *knowledge* of established cause-effect relationships that an intelligent cause operated in the past to produce the specified complexity or information content necessary to the origin of life.

Thus, we do not infer design out of ignorance but because of what we know about the demonstrated causal powers of natural entities and agency, respectively. We infer design using the standard uniformitarian method of reasoning employed in all historical sciences. These inferences are no more based upon ignorance than well-grounded inferences in geology, archeology, or paleontology are—where provisional knowledge of cause-effect relationships derived from present experience guides inferences about the causal past. Recent developments in the information sciences merely help formalize knowledge of these relationships, allowing us to make inferences about the causal histories of various artifacts, entities, or events based upon the complexity and information-theoretic signatures they exhibit.[120] In any case, present knowledge of established cause-effect relationships, not ignorance, justifies the design inference as the best explanation for the origin of specified complexity in both physics and biology.

5.3 Intelligent Design: A *Vera Causa*?

Of course, many would admit that both biological organisms and the contingent features of physical laws manifest complexity and specificity. Nevertheless, they would argue that we cannot infer intelligent design from the presence of complexity and specificity in objects that antedate the origin of human beings. Such critics argue that we *may* justifiably infer a past human intelligence operating (within the scope of human history) from a specified and complex (that is, information-rich) artifact or event, but only because we already know that intelligent human agents exist. But, such critics argue, since we do not know whether an intelligent agent existed prior to humans, inferring the action of a designing mind antedating the advent of humans cannot be justified, even if we know of specified information-rich effects that clearly preexist the origin of human beings.

Note, however, that many well-accepted design inferences do not depend on prior knowledge of a designing intelligence in close spatial or temporal proximity to the effect in question. SETI researchers, for example, do not already know whether an extraterrestrial intelligence exists. Yet they assume that the presence of a large amount of specified complexity (such as the first one hundred prime numbers in sequence) would establish the existence of one. SETI seeks precisely to establish the existence of other intelligences in an unknown domain. Similarly, anthropologists have often revised their estimates for the beginning of human history or civilization because they discovered complex and (functionally) specified artifacts dating from times that antedated their previous estimates for the origin of *Homo sapiens*. Most inferences to design establish the existence or activity of a mental agent operating in a time or place where the presence of such agency was previously unknown. Thus, inferring the activity of a designing intelligence from a time prior to the advent of humans on earth does not have a qualitatively different epistemological status than other design inferences that critics already accept as legitimate.

Yet some would still insist that we cannot legitimately postulate such an agent as an explanation for the origin of specified complexity in life since living systems, as organisms rather than simple machines, far exceed the complexity of systems designed by human agents. Thus, such critics argue, invoking an intelligence similar to that which humans possess would not suffice to explain the exquisite complexity of design present in biological systems. To explain that degree of complexity would require a "superintellect" (to use Fred Hoyle's phrase). Yet, since we have no experience or knowledge of such a super-intelligence, we cannot invoke one as a possible cause for the origin of life. Indeed, we have no knowledge of the causal powers of such a hypothetical agent.

This objection derives from the so-called *vera causa* principle—an important methodological guideline in the historical sciences. The *vera causa* principle asserts that historical scientists seeking to explain an event in the distant past (such as the origin of life) should postulate (or prefer in their postulations) only causes that are sufficient to produce the effect in question and that are known to exist by observation in the present.[121] Darwin, for example, marshaled this methodological consideration as a reason for preferring his theory of natural selection over special creation. Scientists, he argued, can observe natural selection producing biological change; they cannot observe God creating new species.[122]

Even so, Darwin admitted that he could not observe natural selection creating the kind of large-scale morphological changes that his theory required. For this reason, he had to extrapolate beyond the known causal powers of natural selection to explain the origin of morphological novelty during the history of life. Since natural selection was known to produce small-scale changes in a short period of time, he reasoned that it might plausibly produce large-scale changes over vast periods of time.[123] Historical scientists have long regarded such extrapolations as a legitimate way of generating possible explanatory hypotheses in accord with the *vera causa* principle.[124]

Yet, if one admits such reasoning as a legitimate extension of the *vera causa* principle for evolutionary arguments, it seems difficult to exclude consideration of the design hypothesis—indeed, even a theistic design hypothesis—using the same logic. Humans do have knowledge of intelligent agency as causal entity. Moreover, intelligent agents do have the ability to produce specified complexity. Thus, intelligent agency qualifies as a known cause with known causal powers sufficient to produce a specific effect (namely, specified complexity in DNA) in need of explanation. Granted we do not have direct knowledge of a nonhuman intelligence (at least, not of one with capacities greater than our own) operating in the remote past. Yet neither did Darwin have direct knowledge of natural selection operating in the past, nor did he have direct knowledge of natural selection producing large-scale morphological changes in the present. Instead, Darwin postulated a cause for the origin of morphological innovation that resembled one that he could observe in the present but which exceeded what he could observe about that cause in the magnitude of its efficacy. Using a similar logic, one might postulate the past activity of an intelligence similar to human intelligence in its rationality but greater than human intelligence in its design-capacity in order to explain the extreme sophistication and complexity of design present in biological systems. Such a postulation would, like Darwin's, constitute an extrapolation from what we know directly about the powers of a causal entity, in this case, human intelligence. But it would not violate the *vera causa* principle any more than Darwin's extrapolation did.

5.4 But Is It Science?

Of course, many simply refuse to consider the design hypothesis on the grounds that it does not qualify as "scientific". Such critics affirm an extraevidential principle known as "methodological naturalism".[125] Methodological naturalism (MN) asserts that for a hypothesis, theory, or explanation to qualify as "scientific" it must invoke only naturalistic or materialistic causes. Clearly, on this definition, the design hypothesis does not qualify as "scientific". Yet, even if one grants this definition, it does not follow that some nonscientific (as defined by MN) or metaphysical hypothesis may not constitute a better, more causally adequate, explanation. Indeed, this essay has argued that, whatever its classification, the design hypothesis does constitute a better explanation than its naturalistic rivals for the origin of specified complexity in both physics and biology. Surely, simply classifying this argument as metaphysical does not refute it. In any case, methodological naturalism now lacks a compelling justification as a normative definition of science. First, attempts to justify methodological naturalism by reference to metaphysically neutral (that is, non-question begging) demarcation criteria have failed.[126] (See Appendix, pp. 151–211, "The Scientific Status of Intelligent Design".) Second, asserting methodological naturalism as a normative principle for all of science has a negative effect on the practice of certain scientific disciplines. In origin-of-life research, methodological naturalism artificially restricts inquiry and prevents scientists from seeking the most truthful, best, or even most empirically adequate explanation. The question that must be asked about the origin of life is not "Which materialistic scenario seems most adequate?" but "What actually caused life to arise on earth?" Clearly, one of the possible answers to this latter question is "Life was designed by an intelligent agent that existed before the advent of humans." Yet if one accepts methodological naturalism as normative, scientists may not consider this possibly true causal hypothesis. Such an exclusionary logic diminishes the claim to theoretical superiority for any remaining hypothesis and raises the possibility that the best "scientific" explanation (as defined by methodological naturalism) may not, in fact, be the best. As many historians and philosophers of science now recognize, evaluating scientific theories is an inherently comparative enterprise. Theories that gain acceptance in artificially constrained competitions can claim to be neither "most probably true" nor "most empirically adequate". Instead, at best they can achieve the status of the "most probably true or adequate among an artificially limited set of options". Openness to the design hypothesis, therefore, seems necessary to a fully rational historical biology —that is, to one that seeks the truth "no holds barred".[127]

5.5 Conclusion

For almost 150 years many scientists have insisted that "chance and necessity"—happenstance and law—jointly suffice to explain the origin of life and the features or the universe necessary to sustain it. We now find, however, that materialistic thinking—with its reliance upon chance and necessity—has failed to explain the specificity and complexity of both the contingent features of physical law and the biomacromolecules upon which life depends. Even so, many scientists insist that to consider another possibility would constitute a departure from both science and reason itself. Yet ordinary reason, and much scientific reasoning that passes under the scrutiny of materialist sanction, not only recognizes but requires us to recognize the causal activity of intelligent agents. The sculptures of Michelangelo, the software of the Microsoft corporation, the inscribed steles of Assyrian kings—each bespeaks the prior action of intelligent agents. Indeed, everywhere in our high-tech environment, we observe complex events, artifacts, and systems that impel our minds to recognize the activity of other minds —minds that communicate, plan, and design. But to detect the presence of mind, to detect the activity of intelligence in the echo of its effects, requires a mode of reasoning—indeed, a form of knowledge—the existence of which science, or at least official biology, has long excluded. Yet recent developments in the information sciences now suggest a way to rehabilitate this lost way of knowing. Perhaps, more importantly, evidence from biology and physics now strongly suggests that mind, not just matter, played an important role in the origin of our universe and in the origin of the life that it contains.

NOTES

¹ W. A. Dembski, *The Design Inference: Eliminating Chance through Small Probabilities*, Cambridge Studies in Probability, Induction, and Decision Theory (Cambridge: Cambridge University Press, 1998), pp. 1–35.

² The term information content is variously used to denote both specified complexity and unspecified complexity. This essay will use the term to denote the presence of both complexity and specificity. This will prove important to the argument of this essay because a sequence of symbols that is merely complex but not specified (such as "wnsgdtej3dmzcknvcnpd") does not, on Dembski's theory, necessarily indicate the activity of a designing intelligence. Thus, it might be argued that design arguments based upon the presence of information content commit a fallacy of equivocation by inferring design from a type of "information" (i.e., unspecified information) that could result from random natural processes. One can eliminate this ambiguity, however, by defining information content as equivalent to the joint properties of complexity and specification. Though the term is not used this way universally in classical information theory (where the term information can refer to mere improbability or complexity without necessarily implying specificity), biologists have used it to denote both complexity and specificity since the late 1950s. Indeed, Francis Crick and others equated "biological information" not only with complexity but also with what they called "specificity"—where they understood specificity to mean "necessary to function" (A. Sarkar, "Biological Information: A Skeptical Look at Some Central Dogmas of Molecular Biology", in *The Philosophy and History of Molecular Biology: New Perspectives*, ed. S. Sarkar, Boston Studies in the Philosophy of Science [Dordrecht, Netherlands, 1996], 191).

³ Specified complexity is not isomorphic, however, with the terms "Shannon information", "information carrying capacity", or "syntactic information", as defined by classical information theory. Though Shannon's theory and equations provided a powerful way to measure the amount of information that could be transmitted across a communication channel, it had important limits. In particular, it did not, and could not, distinguish merely improbable sequences of symbols from those that conveyed a message or a functionally specified sequence. As Warren Weaver made clear in 1949, "the word information in this theory is used in a special mathematical sense that must not be confused with its ordinary usage. In particular, information must not be confused with meaning" (C. E. Shannon and W. Weaver, *The Mathematical Theory of Communication* [Urbana, Ill.: University of Illinois Press, 1949], p. 8). Information theory could measure the "information-carrying capacity" or the "syntactic information" of a given sequence of symbols but could not dis-

tinguish the presence of a meaningful or functional arrangement of symbols from a random sequence (e.g., "we hold these truths to be self-evident" v. "ntnyhiznlhteqkhgdsjh"). Thus, Shannon information theory could quantify the amount of functional or meaningful information that *might be present* in a given sequence of symbols or characters, but it could not distinguish the status of a functional or message-bearing text from random gibberish. In essence, therefore, Shannon's theory provided a measure of complexity or improbability but remained silent upon the important question of whether a sequence of symbols is functionally specified or meaningful.

[4] K. Giberson, "The Anthropic Principle", *Journal of Interdisciplinary Studies* 9 (1997): 63–90, and response by Steven Yates, pp. 91–104.

[5] P. Davies, *The Cosmic Blueprint* (New York: Simon and Schuster, 1988), p. 203.

[6] G. Greenstein, *The Symbiotic Universe: Life and Mind in the Cosmos* (New York: Morrow, 1988), pp. 26–27.

[7] Greenstein himself does not favor the design hypothesis. Instead, he favors the so-called "participatory universe principle" or "PAP". PAP attributes the apparent design of the fine tuning of the physical constants to the universe's (alleged) need to be observed in order to exist. As he says, the universe "brought forth life in order to exist . . . that the very Cosmos does not exist unless observed" (ibid., p. 223).

[8] F. Hoyle, "The Universe: Past and Present Reflections", *Annual Review of Astronomy and Astrophysics* 20 (1982): 16.

[9] W. Craig, "Cosmos and Creator", *Origins & Design* 20, no. 2 (spring 1996): 23.

[10] J. Leslie, "Anthropic Principle, World Ensemble, Design", *American Philosophical Quarterly* 19, no. 2 (1982): 150.

[11] P. Davies, *The Superforce: The Search for a Grand Unified Theory of Nature* (New York: Simon and Schuster, 1984), p. 243.

[12] D. Halliday, R. Resnick, and G. Walker, *Fundamentals of Physics*, 5th ed. (New York: John Wiley and Sons, 1997), p. A23.

[13] J. Barrow and F. Tipler, *The Anthropic Cosmological Principle* (Oxford: Oxford University Press, 1986), pp. 295–356, 384–444, 510–56; J. Gribbin and M. Rees, *Cosmic Coincidences* (London: Black Swan, 1991), pp. 3–29, 241–69; H. Ross, "The Big Bang Model Refined by Fire", in W. A. Dembski, ed., *Mere Creation: Science, Faith and Intelligent Design* (Downers Grove, Ill.: InterVarsity Press, 1998), pp. 372–81.

[14] A. Guth and M. Sher, "Inflationary Universe: A Possible Solution to the Horizon and Flatness Problems", *Physical Review* 23, no. 2 (1981): 348.

[15] For those unfamiliar with exponential notation, the number 10^{60} is the same as 10 multiplied by itself 60 times or 1 with 60 zeros written after it.

[16] P. Davies, *God and the New Physics* (New York: Simon and Schuster, 1983), p. 188.

[17] R. Penrose, *The Emperor's New Mind* (New York: Oxford, 1989), p. 344.

[18] A. Linde, "The Self–Reproducing Inflationary Universe", *Scientific American* 271 (November 1994): 48–55.

[19] C. Longley, "Focusing on Theism", *London Times*, January 21, 1989, p. 10.

[20] W. Craig, "Barrow and Tipler on the Anthropic Principle v. Divine Design", *British Journal for the Philosophy of Science* 38 (1988): 389–95.

[21] J. W. Richards, "Many Worlds Hypotheses: A Naturalistic Alternative to Design", *Perspectives on Science and Christian Belief* 49, no. 4 (1997): 218–27.

[22] R. Collins, "The Fine-Tuning Design Argument: A Scientific Argument for the Existence of God", in M. Murray, ed., *Reason for the Hope Within* (Grand Rapids, Mich.: Eerdmans, 1999), p. 61.

[23] Ibid., pp. 60–61.

[24] Craig, "Cosmos", p. 24.

[25] R. Swinburne, "Argument from the Fine Tuning of Universe", in J. Leslie, ed., *Physical Cosmology and Philosophy* (New York: Macmillan, 1990), pp. 154–73.

[26] Originally the many-worlds hypothesis was proposed for strictly scientific reasons as a solution to the so-called quantum-measurement problem in physics. Though its efficacy as an explanation within quantum physics remains controversial among physicists, its use there does have an empirical basis. More recently, however, it has been employed to serve as an alternative non-theistic explanation for the fine tuning of the physical constants. This use of the MWH does seem to betray a metaphysical desperation.

[27] Longley, "Focusing", p. 10.

[28] J. Polkinghorne, "So Finely Tuned a Universe of Atoms, Stars, Quanta & God", *Commonweal*, August 16, 1996, p. 16.

[29] B. Alberts, "The Cell as a Collection of Protein Machines: Preparing the Next Generation of Molecular Biologists", *Cell* 92 (February 8, 1998): 291.

[30] M. Behe, *Darwin's Black Box* (New York: Free Press, 1996), pp. 51–73.

[31] According to the neo-Darwinian theory of evolution, organisms evolved by natural selection acting on random genetic mutations. If these genetic mutations help the organism to survive better, they will be preserved in subsequent generations, while those without the mutation will die off faster. For instance, a Darwinian might hypothesize that giraffes born with longer necks were able to reach the leaves of trees more easily, and so had greater survival rates, than giraffes with shorter necks. With time, the necks of giraffes grew longer and longer in a step-by-step process because natural selection favored longer necks. But the intricate machine-like systems in the cell could not have been selected in such a step-by-step process, because not every step in the assembly of a molecular machine enables the cell to survive better. Only when the molecular machine is fully assembled can it function and thus enable a cell to survive better than cells that do not have it.

[32] Behe, *Darwin's*, pp. 165–86.

[33] W. A. Dembski, *Intelligent Design: The Bridge between Science and Theology* (Downers Grove, Ill.: InterVarsity Press, 1999), pp. 146–49.

[34] F. Sanger and H. Tuppy, "The Amino Acid Sequence in the Phenylalanyl Chain of Insulin, 1 and 2", *Biochemical Journal* 49 (1951): 463–80; F. Sanger and E. O. P. Thompson, "The Amino Acid Sequence in the Glycyl Chain of Insulin. 1: The Identification of Lower Peptides from Partial Hydrolysates", *Biochemical Journal* 53 (1953): 353–74; H. Judson, *Eighth Day of Creation* (New York: Simon and Schuster, 1979), pp. 213, 229–35, 255–61, 304, 334–35.

[35] J. C. Kendrew, G. Bodo, H. M. Dintzis, R. G. Parrish, and H. Wyckoff, "A Three-Dimensional Model of the Myoglobin Molecule Obtained by X-Ray Analysis", *Nature* 181 (1958): 664–66; Judson, *Eighth Day*, pp. 562–63.

[36] B. Alberts, D. Bray, J. Lewis, M. Raff, K. Roberts, and J. D. Watson, *Molecular Biology of the Cell* (New York: Garland, 1983), pp. 91–141.

[37] Judson, *Eighth Day*, p. 611.

[38] J. Watson and F. Crick, "A Structure for Deoxyribose Nucleic Acid", *Nature* 171 (1953): 737–38.

[39] Ibid.; J. Watson and F. Crick, "Genetical Implications of the Structure of Deoxyribose Nucleic Acid", *Nature* 171 (1953): 964–67.

[40] Judson, *Eighth Day*, pp. 245–46.

[41] Ibid., pp. 335–36.

[42] R. Dawkins, *River out of Eden* (New York: Basic Books, 1995), p. 10.

[43] B. Gates, *The Road Ahead* (Boulder, Col.: Blue Penguin, 1996), p. 228.

[44] For a good summary and critique of different naturalistic models, see especially K. Dose, "The Origin of Life: More Questions than Answers", *Interdisciplinary Science Review* 13, no. 4 (1998): 348–56; H. P. Yockey, *Information Theory and Molecular Biology* (Cambridge: Cambridge University Press, 1992), pp. 259–93; C. Thaxton, W. Bradley, and R. Olsen, *The Mystery of Life's Origin* (Dallas: Lewis and Stanley, 1992); C. Thaxton and W. Bradley, "Information and the Origin of Life", in *The Creation Hypothesis: Scientific Evidence for an Intelligent Designer*, ed. J. P. Moreland (Downers Grove, Ill.: InterVarsity Press, 1994), pp. 173–210; R. Shapiro, *Origins* (London: Heinemann, 1986), pp. 97–189; S. C. Meyer, "The Explanatory Power of Design: DNA and the Origin of Information", in Dembski, *Mere Creation*, pp. 119–34. For a contradictory hypothesis, see S. Kauffman, *The Origins of Order* (Oxford: Oxford University Press, 1993), pp. 287–341.

[45] G. Wald, "The Origin of Life", *Scientific American* 191 (August 1954): 44–53; Shapiro, *Origins*, p. 121.

[46] F. Crick, "The Origin of the Genetic Code", *Journal of Molecular Biology* 38 (1968): 367–79; H. Kamminga, *Studies in the History of Ideas on the Origin of Life* (Ph.D. diss., University of London, 1980), pp. 303–4.

[47] C. de Duve, *Blueprint for a Cell: The Nature and Origin of Life* (Burlington, N.C.: Neil Patterson Publishers, 1991), p. 112; F. Crick, *Life Itself* (New York:

Simon and Schuster, 1981), pp. 89–93; H. Quastler, *The Emergence of Biological Organization* (New Haven: Yale University Press, 1964), p. 7.

[48] H. J. Morowitz, *Energy Flow in Biology* (New York: Academic Press, 1968), pp. 5–12.

[49] F. Hoyle and C. Wickramasinghe, *Evolution from Space* (London: J. M. Dent, 1981), pp. 24–27.

[50] A. G. Cairns–Smith, *The Life Puzzle* (Edinburgh: Oliver and Boyd, 1971), pp. 91–96.

[51] I. Prigogine, G. Nicolis, and A. Babloyantz, "Thermodynamics of Evolution", *Physics Today*, November 1972, p. 23.

[52] Yockey, *Information Theory*, pp. 246–58; H. P. Yockey, "Self Organization, Origin of Life Scenarios and Information Theory", *Journal of Theoretical Biology* 91 (1981): 13–31; see also Shapiro, *Origins*, pp. 117–31.

[53] Prigogine, Nicolis, and Babloyantz, "Thermodynamics", p. 23.

[54] Cairns-Smith, *Life Puzzle*, p. 95.

[55] Alberts et al., *Molecular Biology*, p. 118.

[56] Actually, Sauer counted sequences that folded into stable three-dimensional configurations as functional, though many sequences that fold are not functional. Thus, his results actually underestimate the probabilistic difficulty.

[57] J. Reidhaar-Olson and R. Sauer, "Functionally Acceptable Solutions in Two Alpha-Helical Regions of Lambda Repressor", *Proteins, Structure, Function, and Genetics* 7 (1990): 306–10; J. Bowie and R. Sauer, "Identifying Determinants of Folding and Activity for a Protein of Unknown Sequences: Tolerance to Amino Acid Substitution", *Proceedings of the National Academy of Sciences*, USA 86 (1989): 2152–56; J. Bowie, J. Reidhaar-Olson, W. Lim, and R. Sauer, "Deciphering the Message in Protein Sequences: Tolerance to Amino Acid Substitution", *Science* 247 (1990): 1306–10; M. Behe, "Experimental Support for Regarding Functional Classes of Proteins to Be Highly Isolated from Each Other", in J. Buell and G. Hearns, eds., *Darwinism: Science or Philosophy?* (Dallas: Haughton Publishers, 1994), pp. 60–71; Yockey, *Information Theory*, pp. 246–58.

[58] See also D. Axe, N. Foster, and A. Ferst, "Active Barnase Variants with Completely Random Hydrophobic Cores", *Proceedings of the National Academy of Sciences, USA* 93 (1996): 5590.

[59] W. Dembski, *The Design Inference: Eliminating Chance through Small Probabilities* (Cambridge: Cambridge University Press, 1998), pp. 67–91, 175–223. Dembski's universal probability bound actually reflects the "specificational" resources, not the probabilistic resources, in the universe. Dembski's calculation determines the number of specifications possible in finite time. It nevertheless has the effect of limiting the "probabilistic resources" available to explain the origin of any *specified* event of small probability. Since living systems are precisely specified systems of small probability, the universal probability bound

effectively limits the probabilistic resources available to explain the origin of specified biological information (ibid., 1998, pp. 175–229).

[60] Cassette mutagenesis experiments have usually been performed on proteins of about one hundred amino acids in length. Yet extrapolations from these results can generate reasonable estimates for the improbability of longer protein molecules. For example, Sauer's results on the proteins lambda repressor and arc repressor suggest that, on average, the probability at each site of finding an amino acid that will maintain functional sequencing (or, more accurately, that will produce folding) is less than 1 in 4 (1 in 4.4). Multiplying 1/4 by itself 150 times (for a protein 150 amino acids in length) yields a probability of roughly 1 chance in 10^{91}. For a protein of that length the probability of attaining both exclusive peptide bonding and homochirality is also about 1 chance in 10^{91}. Thus, the probability of achieving all the necessary conditions of function for a protein 150 amino acids in length exceeds 1 chance in 10^{180}.

[61] Dembski, *Design Inference*, pp. 67–91, 175–214; cf. E. Borel, *Probabilities and Life*, trans. M. Baudin (New York: Dover, 1962), p. 28.

[62] E. Pennisi, "Seeking Life's Bare Genetic Necessities", *Science* 272 (1996): 1098–99; A. Mushegian and E. Koonin, "A Minimal Gene Set for Cellular Life Derived by Comparison of Complete Bacterial Genomes", *Proceedings of the National Academy of Sciences, USA* 93 (1996): 10268–73; C. Bult et al., "Complete Genome Sequence of the Methanogenic Archaeon, *Methanococcus Jannaschi*", *Science* 273 (1996): 1058–72.

[63] Dembski, *Design Inference*, pp. 67–91, 175–223.

[64] P. T. Mora, "Urge and Molecular Biology", *Nature* 199 (1963): 212–19.

[65] I. Hacking, *The Logic of Statistical Inference* (Cambridge: Cambridge University Press, 1965), pp. 74–75.

[66] Dembski, *Design Inference*, pp. 47–55.

[67] C. de Duve, "The Beginnings of Life on Earth", *American Scientist* 83 (1995): 437.

[68] H. Quastler, *The Emergence of Biological Organization* (New Haven, Conn.: Yale University Press, 1964), p. 7.

[69] A. I. Oparin, *The Origin of Life*, trans. S. Morgulis (New York: Macmillan Co., 1938), pp. 64–103; S. C. Meyer, *Of Clues and Causes. A Methodological Interpretation of Origin of Life Studies* (Ph.D. diss., Cambridge University, 1991), pp. 174–79, 194–98, 211–12.

[70] Oparin, *Origin*, pp. 107–8.

[71] Ibid., pp. 133–35.

[72] Ibid., pp. 148–59.

[73] Ibid., pp. 195–96.

[74] A. I. Oparin, *Genesis and Evolutionary Development of Life* (New York: Academic Press, 1968), pp. 146–47.

[75] De Duve, *Blueprint*, p. 187.

[76] G. Joyce and L. Orgel, "Prospects for Understanding the Origin of the RNA World", in R. F. Gesteland and J. F. Atkins, eds., *RNA World* (Colds Spring Harbor, N.Y.: Colds Spring Harbor Laboratory Press, 1993), pp. 8–13.

[77] J. von Neumann, *Theory of Self-Reproducing Automata*, ed. and completed by A. Berks (Urbana, Ill.: University of Illinois Press, 1966).

[78] E. Wigner, "The Probability of the Existence of a Self-Reproducing Unit", in *The Logic of Personal Knowledge: Essays Presented to Michael Polanyi on His Seventieth Birthday* (London: Routledge and Paul, 1961), pp. 231–35.

[79] P. T. Landsberg, "Does Quantum Mechanics Exclude Life?" *Nature* 302 (1964): 928–30.

[80] H. J. Morowitz, "The Minimum Size of the Cell", in M. O'Connor and G. Wolstenholme, eds., *Principles of Biomolecular Organization* (London: Churchill, 1966), pp. 446–59; Morowitz, *Energy*, pp. 10–11.

[81] T. Dobzhansky, discussion of G. Schramm's paper, in S. W. Fox, ed., *The Origins of Prebiological Systems and of Their Molecular Matrices* (New York: Academic Press, 1965), p. 310; see also H. H. Pattee, "The Problem of Biological Hierarchy", in C. H. Waddington, ed., *Toward a Theoretical Biology*, vol. 3 (Edinburgh: Edinburgh University Press, 1970), p. 123.

[82] P. T. Mora, "The Folly of Probability", in Fox, *Origins*, pp. 311–12; L. V. Bertalanffy, *Robots, Men and Minds* (New York: George Braziller, 1967), p. 82.

[83] R. Dawkins, *The Blind Watchmaker* (London: Longman, 1986), pp. 47–49.

[84] B. Küppers, "The Prior Probability of the Existence of Life", in L. Krüger, G. Gigerenzer, and M. S. Morgan, eds., *The Probabilistic Revolution* (Cambridge: MIT Press, 1987), pp. 355–69.

[85] Ibid., p. 366.

[86] Ibid.

[87] Dawkins, *Watchmaker*, pp. 47–49.

[88] P. Nelson, "Anatomy of a Still-Born Analogy", *Origins & Design* 17, no. 3 (1996): 12.

[89] De Duve, "Beginnings", p. 437.

[90] Morowitz, *Energy*, pp. 5–12.

[91] G. Steinman and M. N. Cole, "Synthesis of Biologically Pertinent Peptides under Possible Primordial Conditions", *Proceedings of the National Academy of Sciences, USA* 58 (1967): 735–41; G. Steinman, "Sequence Generation in Prebiological Peptide Synthesis", *Archives of Biochemistry and Biophysics* 121 (1967): 533–39; R. A. Kok, J. A. Taylor, and W. L. Bradley, "A Statistical Examination of Self-Ordering of Amino Acids in Proteins", *Origins of Life and Evolution of the Biosphere* 18 (1988): 135–42.

[92] D. Kenyon and G. Steinman, *Biochemical Predestination* (New York: McGraw-Hill, 1969), pp. 199–211, 263–66.

[93] I. Prigogine and G. Nicolis, *Self-Organization in Nonequilibrium Systems* (New York: John Wiley, 1977), pp. 339–53, 429–47.

[94] Kauffman, *Origins*, pp. 285–341.

[95] De Duve, "Beginnings", pp. 428–37; C. de Duve, *Vital Dust: Life as a Cosmic Imperative* (New York: Basic Books, 1995).

[96] D. Kenyon, foreword to C. Thaxton, W. Bradley, and R. Olsen, *The Mystery of Life's Origin* (Dallas: Lewis and Stanley, 1992), pp. v–viii; D. Kenyon and G. Mills, "The RNA World: A Critique", *Origins & Design* 17, no. 1 (1996): 12–16; D. Kenyon and P. W. Davis, *Of Pandas and People: The Central Question of Biological Origins* (Dallas: Haughton, 1993); S. C. Meyer, "A Scopes Trial for the '90's", *The Wall Street Journal*, December 6, 1993, p. A14; Kok, Taylor, and Bradley, "Examination", pp. 135–42.

[97] Steinman and Cole, "Synthesis", pp. 735–41; Steinman, "Sequence Generation", pp. 533–39.

[98] Kok, Taylor, and Bradley, "Examination", pp. 135–42.

[99] Alberts et al., *Molecular Biology*, p. 105.

[100] Küppers, "Prior Probability", p. 364.

[101] Note that the "RNA world" scenario was not devised to explain the origin of the sequence specificity of biomacromolecules. Rather it was proposed as an explanation for the origin of the interdependence of nucleic acids and proteins in the cellular information processing system. In extant cells, building proteins requires instructions from DNA, but information on DNA cannot be processed without many specific proteins and protein complexes. This poses a "chicken-or-egg" dilemma. The discovery that RNA (a nucleic acid) possesses limited catalytic properties (as modern proteins do) suggested a way to split the horns of this dilemma. By proposing an early earth environment in which RNA performed both the enzymatic functions of modern proteins and the information storage function of modern DNA, "RNA first" advocates sought to formulate a scenario making the functional interdependence of DNA and proteins unnecessary to the first living cell. In so doing, they sought to make the origin of life a more tractable problem from a chemical evolutionary point of view. In recent years, however, many problems have emerged with the RNA world (Shapiro, *Origins*, pp. 71–95; Kenyon, "RNA World", pp. 9–16). To name just one, RNA possesses very few of the specific catalytic properties necessary to facilitate the expression of the genetic information on DNA. In any case, by addressing a separate problem, the RNA world presupposed a solution to, but did not solve, the sequence specificity or information problem. A recent article heralding a breakthrough for "RNA world" scenarios makes this clear. After reporting on RNA researcher Jack Szostak's successful synthesis of RNA molecules with some previously unknown catalytic properties, science writer John Horgan makes a candid admission: "Szostak's work leaves a major question unanswered: How did RNA, self-catalyzing or not, arise in the first place?" (J. Horgan, "The World according to RNA", *Scientific American*, January 1996, p. 27; R. Shapiro, "Prebiotic Ribose Synthesis: A Critical Analysis", *Origins of Life and Evolution of the Biosphere* 18 [1988]: 71–95; A. Zaug and T. Cech, "The

Intervening Sequence RNA of Tetrahymena Enzyme", *Science* 231 [1986]: 470–75; T. Cech, "Ribozyme Self-Replication?" *Nature* 339 [1989] 507–8; Kenyon, "RNA World", pp. 9–16.

[102] De Duve, "Beginnings", p. 437.

[103] This, in fact, happens where adenine and thymine do interact chemically in the complementary base pairing *across* the message-bearing axis of the DNA molecule. Recent experiments have also shown that deoxyribonucleoside 5' triphosphates (i.e., the nucleotide bases joined with necessary sugar and phosphate molecules) will form repetitive sequences in solution *even when polymerization is facilitated by a polymerizing enzyme such as DNA polymerase* (N. Ogata and T. Miura, "Genetic Information 'Created' by Archaebacterial DNA Polymerase", *Biochemistry Journal* 324 [1997]: 567–71).

[104] M. Polanyi, "Life's Irreducible Structure", *Science* 160 (1968): 1308–12, esp. 1309.

[105] The information-carrying capacity of any symbol in a sequence is inversely related to the probability of its occurrence. The informational capacity of a sequence as a whole is inversely proportional to the product of the individual probabilities of each member in the sequence. Since chemical affinities between constituents ("symbols") increase the probability of the occurrence of one given another (i.e., necessity increases probability), such affinities decrease the information-carrying capacity of a system in proportion to the strength and relative frequency of such affinities within the system.

[106] F. Dretske, *Knowledge and the Flow of Information* (Cambridge: MIT Press, 1981), p. 12.

[107] Yockey, "Self Organization", p. 18.

[108] Orgel has drawn a similar distinction between either order or the randomness that characterizes inanimate chemistry and what he calls the "specified complexity" of informational biomolecules (A. I. Orgel, *The Origins of Life on Earth* [New York: Macmillan, 1973], pp. 189ff.; Thaxton, *Mystery*, pp. 130ff.).

[109] H. P. Yockey, "A Calculation of the Probability of Spontaneous Biogenesis by Information Theory", *Journal of Theoretical Biology* 67 (1977): 380.

[110] M. Eigen, *Steps toward Life* (Oxford: Oxford University Press, 1992), p. 12.

[111] Quastler, *Emergence*, p. 16.

[112] This qualification means to acknowledge that chance can produce low levels (less than 500 bits) of specified information (see Dembski, *Design Inference*, pp. 175–223).

[113] Meyer, *Clues*, pp. 77–140.

[114] Ibid.; E. Sober, *Reconstructing the Past* (Cambridge: MIT Press, 1988), pp. 4–5; M. Scriven, "Causes, Connections, and Conditions in History", in W. Dray, ed., *Philosophical Analysis and History* (New York: Harper and Row, 1966), pp. 249–50.

[115] T. R. McDonough, *The Search for Extraterrestrial Intelligence: Listening for Life in the Cosmos* (New York: Wiley, 1987).

[116] Dembski, *Design Inference*, pp. xi–xiii, 1–35.

[117] Ibid., pp. 1–35, 36–66.

[118] P. Lipton, *Inference to the Best Explanation* (New York: Routledge, 1991), pp. 32–88.

[119] Ibid.; S. C. Meyer, "The Methodological Equivalence of Design and Descent: Can There Be a Scientific Theory of Creation?" in J. P. Moreland, ed., *The Creation Hypothesis: Scientific Evidence for an Intelligent Designer* (Downers Grove, Ill.: InterVarsity Press, 1994), pp. 67–112; E. Sober, *The Philosophy of Biology* (San Francisco: Westview Press, 1993), p. 44; Meyer, *Clues*, pp. 77–140.

[120] Dembski, *Design Inference*, pp. 36–66, esp. p. 37.

[121] M. J. S. Hodge, "The Structure and Strategy of Darwin's 'Long Argument' ", *British Journal for the History of Science* 10 (1977): 239.

[122] V. Kavalovski, *The Vera Causa Principle: A Historico-Philosophical Study of a Meta-theoretical Concept from Newton through Darwin* (Ph.D. diss., University of Chicago, 1974), p. 104.

[123] Meyer, *Clues*, pp. 87–91.

[124] Kavalovski, *Vera Causa*, p. 67.

[125] M. Ruse, "Witness Testimony Sheet: McLean v. Arkansas", in M. Ruse, ed., *But Is It Science?* (Buffalo, N.Y.: Prometheus Books, 1988), p. 301; R. Lewontin, "Billions and Billions of Demons", *The New York Review of Books*, January 9, 1997, p. 31; Meyer, "Equivalence", pp. 69–71.

[126] Meyer, "Laws", pp. 29–40; Meyer, "Equivalence", pp. 67–112; S. C. Meyer, "Demarcation and Design: The Nature of Historical Reasoning", in Jitse van der Meer, ed., *Facets of Faith and Science*, vol. 4, *Interpreting God's Action in the World* (Lanham, Md.: University Press of America, 1996), pp. 91–130; L. Laudan, "The Demise of the Demarcation Problem", in Ruse, *Science?*, pp. 337–50; L. Laudan, "Science at the Bar—Causes for Concern", in Ruse, *Science?*, pp. 351–55; A. Plantinga, "Methodological Naturalism", *Origins & Design* 18, no. 1 (1997): 18–27, and no. 2 (1997): 22–34.

[127] P. W. Bridgman, *Reflections of a Physicist*, 2d ed. (New York: Philosophical Library, 1955), p. 535.

MICHAEL J. BEHE

EVIDENCE FOR DESIGN AT
THE FOUNDATION OF LIFE

Urea and Purpose

In the year 1828 the German chemist Friedrich Wöhler heated
ammonium cyanate in his laboratory and was amazed to see
that urea was produced. Why was he amazed? Because ammonium cyanate is an inorganic chemical—one that does not
occur in living organisms. But urea was known to be a biological waste product. Wöhler was the first to demonstrate that a
nonliving substance could give rise to a substance produced by
living organisms. His experiment shattered the distinction between life and nonlife that was thought to exist up until that
time. Moreover, it opened up all of life for scientific study.
For if life is made of ordinary matter, the same as rocks and so
on, then science can study it. And in the more than 170 years
since Wöhler's experiment, science has learned a lot about life.
We have discovered the structure of DNA, cracked the genetic
code, learned to clone genes, and cells, and even whole organisms.

What has the progress of science told us about the ultimate
nature of the universe and life? Well, of course, there are a
lot of opinions on the subject, but I think we can break them
down into two opposite sides. The first side can perhaps be
represented by Richard Dawkins, professor of the public understanding of science at Oxford University. Professor Dawkins
has stated that: "The universe we observe has precisely the
properties we should expect if there is at bottom no design,
no purpose, no evil and no good, nothing but pointless indif-

ference."[1] Certainly a dreary view, but a seriously proposed one.

The second point of view can be represented by Joseph Cardinal Ratzinger, an advisor to Pope John Paul II. About ten years ago Cardinal Ratzinger wrote a little book entitled *In the Beginning: A Catholic Understanding of the Story of Creation and the Fall*. In the book Cardinal Ratzinger wrote:

> Let us go directly to the question of evolution and its mechanisms. Microbiology and biochemistry have brought revolutionary insights here. . . . It is the affair of the natural sciences to explain how the tree of life in particular continues to grow and how new branches shoot out from it. This is not a matter for faith. But we must have the audacity to say that the great projects of the living creation are not the products of chance and error. . . . [They] point to a creating Reason and show us a creating Intelligence, and they do so more luminously and radiantly today than ever before. Thus we can say today with a new certitude and joyousness that the human being is indeed a divine project, which only the creating Intelligence was strong and great and audacious enough to conceive of. Human beings are not a mistake but something willed.[2]

I would like to make three points about the Cardinal's argument. First, unlike Professor Dawkins, Ratzinger says that nature does appear to exhibit purpose and design. Secondly, to support the argument he points to *physical evidence*—the "great products of the living creation", which "point to a creating Reason". Not to philosophical, or theological, or scriptural arguments, but to tangible structures. Thirdly, Ratzinger cites the science of biochemistry—the study of the molecular foundation of life—as having particular relevance to his conclusion. It is my purpose in this essay to show why I think Cardinal Ratzinger has the stronger position, and why Professor Dawkins need not despair.

Explaining the Eye

Of course much of this discussion about the nature of life began in 1859, when Charles Darwin published *The Origin of Species*. In his book Darwin proposed to do what no one had been able to do before him—explain how the great variety and complexity of life might have arisen solely through unguided natural processes. His proposed mechanism was, of course, natural selection acting on random variation. In a nutshell, Darwin recognized that there is variety in all species. Some members of a species are larger than others, some faster, some darker in color. Darwin knew that not all members of a species that are born will survive to reproduce, simply because there is not enough food to sustain them all. And so he reasoned that the ones whose chance variation gave them an edge in the struggle to survive would tend to survive and leave offspring. If the variation could be inherited, then over time the characteristics of the species might change. And over great periods of time, great changes might occur.

Darwin's theory was a very elegant idea. Nonetheless, even in the mid-nineteenth century biologists knew of a number of biological systems that did not appear to be able to be built in the gradual way that Darwin envisioned. One in particular was the eye. Biologists of the time knew that the eye was a very complex structure, containing many components, such as a lens, retina, tear ducts, ocular muscles, and so forth. They knew that if an animal were so unfortunate as to be born without one of the components, the result would be a severe loss of vision or outright blindness. So they doubted that such a system could be put together in the many steps required by natural selection.

Charles Darwin, however, knew about the eye too. And he wrote about it in a section of the *Origin of Species* appropriately entitled "Organs of Extreme Perfection and Complication", in which he said that he did not really know how the eye might have evolved. Nonetheless, he wrote that if you look at the eyes of modern organisms, you see considerable variety. In some organisms there really is not an "eye", but rather just a patch of light-sensitive cells. Now, that arrangement is sufficient for enabling an organism to know if it is in light or darkness, but it does not enable an organism to determine which direction the light is coming from, because light coming from virtually any angle will stimulate the light-sensitive cells. However, Darwin continued, if you take that patch of light-sensitive cells and place it in a small depression, as is seen in some modern animals, light coming from one side will cast a shadow over part of the light-sensitive spot, while the rest is illuminated. In theory such an arrangement could allow the creature to determine which direction the light is coming from. And that would be an improvement. If the cup were deepened, the direction-finding ability would be increased. And if the cup were filled with a gelatinous material, that could be the beginning of a crude lens, a further improvement. Using arguments like these, Darwin was able to convince many of his contemporaries that a gradual evolutionary pathway led from something as simple as a light-sensitive spot to something as complicated as the modern vertebrate eye. And if evolution could explain the eye . . . well, what could it not explain?

But there was a question left unaddressed by Darwin's scheme—where did the light-sensitive spot come from? It seems an odd starting point, since most objects are not light sensitive. Nonetheless, Darwin decided not even to attempt to address the question. He wrote that: "How a nerve comes to be sensitive to light hardly concerns us more than how life itself originated."[3]

Well, in the past half-century science has become interested in both those questions: the mechanism of vision and the origin of life. Nonetheless, Darwin was correct, I think, to refuse to address the question, because the science of his day did not have the physical or conceptual tools to begin to investigate it. Just to get a flavor of the science of the mid-nineteenth century, remember that atoms—the basis of all chemistry—were then considered to be theoretical entities. No one was sure if they really existed. The cell, which we now know to be the basis of life, was thought to be a simple glob of protoplasm, not much more than a microscopic piece of Jell-O. So Darwin refused to address the question and left it as a black box in the hope that future discoveries would vindicate his theory.

"Black box" is a phrase used in science to indicate some machine or system that does something interesting, but no one knows how it works. Its mechanism is unknown because we cannot see inside the box to observe it, or if we can see the workings, they are so complicated that we still do not understand what is going on. For most of us (and certainly for me) a good example of a black box is a computer. I use a computer to process words or play games, but I do not have the foggiest idea how it works. And even if I were to remove the cover and see the inside circuitry, I still could not say how it worked. Well, to scientists of Darwin's day, the cell was a black box. It did very interesting things, but no one knew how.

When people see a black box in action, they have a psychological tendency to assume that it must be operating by some simple mechanism—the insides of the box must be uncomplicated and working on some easily understood principle. A good example of this tendency was the belief in the spontaneous generation of cellular life from sea mud. In the nineteenth century two prominent scientists and admirers of Darwin—Ernst Haeckel and Thomas Huxley—thought that some mud scraped up by an exploring vessel might be living cells. They could believe this because they thought a cell was, in Haeckel's words, a "simple little lump of albuminous combination of carbon".[4] With the tremendous progress biology has made in this century, of course, we know differently. Now that modern science has opened the black box of the cell, we need to readdress the question that stumped Darwin. What is needed to make a light-sensitive spot? What happens when a photon of light impinges upon a retina?

When a photon first hits the retina, it interacts with a small organic molecule called 11-cis-retinal.[5] The shape of retinal is rather bent, but when retinal interacts with the photon, it straightens out, isomerizing into trans-retinal. This is the signal that sets in motion a whole cascade of events resulting in vision. When retinal changes shape, it forces a change in the shape of the protein rhodopsin, which is bound to it. The change in rhodopsin's shape exposes a binding site that allows the protein transducin to stick to it. Now part of the transducin complex dissociates and interacts with a protein called phosphodiesterase. When that happens, the phosphodiesterase acquires the ability chemically to cut a small organic molecule called cyclic-GMP, turning it into 5'-GMP. There is a lot of cyclic-GMP in the cell, and some of it sticks to another protein called an ion channel. Normally the ion channel allows sodium ions into the cell. When the concentration of cyclic-GMP decreases because of the action of the phosphodiesterase, however, the cyclic-GMP bound to the ion channel eventually falls off, causing a change in shape that shuts the channel. As a result, sodium ions can no longer enter the cell, the concentration of sodium in the cell decreases, and the voltage across the cell membrane changes. That in turn causes a wave of electrical polarization to be sent down the optic nerve to the brain. And, when interpreted by the brain, that is vision. So, this is what modern science has discovered about how Darwin's "simple" light-sensitive spot functions.

Darwin's Criterion

Although most people will surely think the above description of the visual cascade is complicated, it is really just a little sketch of the chemistry of vision that ignores a number of things that a functioning visual system actually requires. For instance, I have not discussed the regeneration of the system —how it gets back to the starting point in preparation for the next incoming photon. Nonetheless, I think that the discussion above is sufficient to show that what Darwin and his contemporaries took as simple starting points have turned out to be enormously complex—much more complex than Darwin ever envisioned.

But how can we tell if the eye and other organisms are too complex to be explained by Darwin's theory? It turns out that Darwin himself gave us a criterion by which to judge his theory. He wrote in the *Origin of Species* that: "If it could be demonstrated that any complex organ existed which could not possibly have been formed by numerous, successive, slight modifications, my theory would absolutely break down."[6] But what sort of organ or system could not be formed by "numerous, successive, slight modifications"? Well, to begin with, one that is *irreducibly complex.* "Irreducibly complex" is a fancy phrase, but it stands for a very simple concept. As I wrote in *Darwin's Black Box: The Biochemical Challenge to Evolution,* an irreducibly complex system is: "a single system which is composed of several well-matched, interacting parts that contribute to the basic function, and where the removal of any one of the parts causes the system to effectively cease functioning."[7] Less formally, the phrase "irreducibly complex" just means that a system has a number of components that interact with each other, and if any are taken away the system no longer works. A good illustration of an irreducibly complex system from our everyday world is a simple mechanical mousetrap. The mousetraps that one buys at the hardware store generally have a wooden platform to which all the other parts are attached. It also has a spring with extended ends, one of which presses against the platform, the other against a metal part called the hammer, which actually does the job of squashing the mouse. When one presses the hammer down, it has to be stabilized in that position until the mouse comes along, and that is the job of the holding bar. The end of the holding bar itself has to be stabilized, so it is placed into a metal piece called the catch. All of these pieces are held together by assorted staples.

Now, if the mousetrap is missing the spring, or hammer, or platform, it does not catch mice half as well as it used to, or even a quarter as well. It does not catch mice at all. Therefore it is irreducibly complex. It turns out that irreducibly complex systems are headaches for Darwinian theory, because they are resistant to being produced in the gradual, step-by-step manner that Darwin envisioned. For example, if we wanted to evolve a mousetrap, where would we start? Could we start with just the platform and hope to catch a few mice rather inefficiently? Then add the holding bar, and improve the efficiency a bit? Then add the other pieces one at a time, steadily improving the whole apparatus? No, of course we cannot do that, because the mousetrap does not work at all until it is essentially completely assembled.

Biochemical Challenges to Darwinism

Mousetraps are one thing, biological systems another. What we really want to know is whether there are any irreducibly complex biological systems, or cellular systems, or biochemical systems. And it turns out that there are many such irreducibly complex systems. Let us consider two examples. The first is called the cilium. A cilium is a little hairlike organelle on the surface of many types of cells. It has the intriguing ability to beat back and forth, moving liquid over the surface of a cell. In some tissue in the lungs, each cell contains hundreds of cilia that beat in synchrony. Interspersed among the ciliated cells are larger ones called goblet cells. The goblet cells secrete mucus into the lining of the lungs, which is swept by the ciliary beating up to the throat where it can be coughed out, along with any dust particles or other foreign objects that might have made their way into the lungs. But what makes a little hairlike organelle beat back and forth? Work in the past several decades has shown that cilia are actually very complicated molecular machines.

The basic structure of a cilium consists of nine double microtubules.[8] [See figure on p. 122.] Each of the double microtubules contains two rings made up of ten and thirteen strands respectively of the protein tubulin. In the middle of the cilium are two single microtubules. All of the microtubules are connected to each other by various types of connectors. Neighboring double microtubules are connected by a protein called nexin. The outer double microtubules are connected to the inner single microtubules by radial spokes. And the two inner microtubules are attached by a small connecting bridge. Additionally, on each double microtubule there are two appendages: an outer dynein bridge and an inner dynein bridge. Although this all sounds complicated, such a brief description cannot do justice to the full complexity of the cilium, which, thorough biochemical studies have shown, contains about two hundred different kinds of protein parts.

But how does the cilium work? Studies have shown that it works by a "sliding-fiber mechanism". Neighboring microtubules are the fibers; dynein is a "motor protein". When the cilium is working, the dynein, bound to one strand, reaches over, attaches to a neighboring microtubule, and pushes down. When that happens, the microtubules start to slide with respect to each other. They would continue to slide until they fell apart, except that they are held together by the linker protein nexin. Initially rather loose, as the fibers slide, the nexin becomes more and more taut. As the tension on the nexin and microtubules increases beyond a certain point, the microtubules bend. Thus the sliding motion is converted into a bending motion.

If one thinks about it, it is easy to see that the cilium is irreducibly complex. If it were not for the microtubules, there would be nothing to slide. If the dynein were missing, the whole apparatus would lie stiff and motionless. And if the nexin linkers were missing, the whole apparatus would fall apart when the dynein started to push the microtubules, as it does in experiments when the nexin linkers are removed.

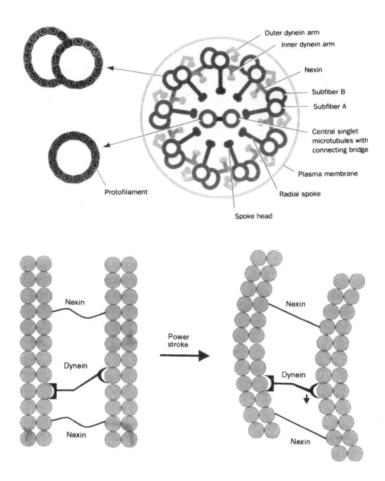

(*Top*) Cross-section of a cilium showing the fused double-ring structure of the outer microtubules, the single-ring structure of the central microtubules, connecting proteins, and dynein motor. (*Bottom*) The sliding motion induced by dynein "walking" up a neighboring microtubule is converted to a bending motion by the flexible linker protein nexin. From Voet and Voet, *Biochemistry*, 2d ed. © 1995 John Wiley and Sons, Inc. Reprinted by permission of John Wiley and Sons, Inc.

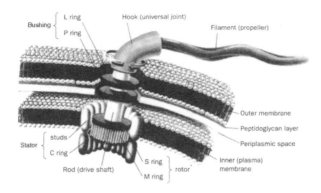

Drawing of a bacterial flagellum showing the filament, hook, and the motor imbedded in the inner and outer cell membranes and the cell wall. From Voet and Voet, *Biochemistry*, 2d ed. © 1995 John Wiley and Sons, Inc. Reprinted by permission of John Wiley and Sons, Inc.

Much like a mousetrap, a cilium needs a number of parts to function. And, again like a mousetrap, its gradual production in a step-by-step Darwinian fashion is quite difficult to envision.

Another example of an irreducibly complex biochemical system is in some ways like the cilium in that it is an organelle for motion. But in other ways it is completely different. The bacterial flagellum is quite literally an outboard motor that enables some bacteria to swim.[2] [See figure above.] Like the machines that power our motorboats, the flagellum is a rotary device, in which the rotating surface pushes against the liquid medium, propelling the bacterium along. The part of the flagellum that acts as the propeller is a long whip-like structure made of a protein called flagellin. The propeller is attached to the drive shaft by hook protein, which acts as a universal joint, allowing freedom of rotation for the propeller and drive shaft. The drive shaft is attached to the rotary motor, which uses a flow of acid from outside of the bacterium to the inside in order to power its turning. The drive shaft has to poke through the bacterial membrane, and several types of proteins act as bushing material to allow that to happen. Although this description makes the flagellum sound complicated, it really does not do justice to its full complexity. Thorough genetic studies have shown that about forty different proteins are required for a functional flagellum, either as parts of the flagellum itself or as parts of the system that builds this machine in the cell. And in the absence of most of those proteins, one does not get a flagellum that spins half as fast as it used to, or a quarter as fast. Either no flagellum gets produced at all, or one that does not work at all. Much like a cilium or mousetrap, the flagellum requires a number of parts to work. Therefore it is irreducibly complex, and its origin presents quite a stumbling block to Darwinian theory.

Darwinian Imagination

I did not discover the cilium or flagellum. It was not I who worked out their mechanisms of action. That work was done by dozens and dozens of laboratories around the world over the course of decades. But if these structures cannot be explained by Darwinian theory, as I contend, then what have other scientists been saying about the origin of molecular machines? One place to look for an answer to that question is in the *Journal of Molecular Evolution*. As its name implies, *JME* was set up specifically to investigate how life might have arisen and then diversified at the molecular level. It is a good journal, which publishes interesting, rigorous material. Of the approximately forty scientists on its editorial board, about fifteen or so are members of the National Academy of Sciences. However, if you pick up a recent copy, you will find that the great majority of papers concern something called "sequence analysis". Briefly, proteins—the components of molecular machines— are made up of "sequences" of amino acids stitched together. Now, if one knows the sequence of amino acids in a protein (or in its gene) then one can compare the sequence to a similar protein from another species and see where the two sequences are the same, similar, or different. For example, suppose one compared the sequence of the oxygen-carrying protein hemoglobin from a dog to that from a horse. One could then ask, are the amino acid residues in the first position of the two proteins the same or different? How about the second position? the third? the fortieth? And so on. Knowing the answer to that question would be interesting and could indicate how closely related the two species are, and that would be an interesting thing to know.

For our purposes, however, the important point to keep in mind is that comparing sequences does not allow one to conclude how complex molecular machines, such as the cilium or flagellum, could have arisen step by Darwinian step. Perhaps an example would help to show why. Suppose that you compared the bones in the forelimb of a dog to those in the forelimb of a horse. And you observed that there were the same number of bones, and they were arranged in a similar pattern. Knowing that would be interesting, and that might allow you to conclude how closely related the animals are, which again would be an interesting thing to know. However, comparing the bones in the forelimb of a dog to those of a horse will not tell you where bones came from in the first place. In order to do that, you have to build models, do experiments, and so forth. It turns out that virtually none of the papers in the *Journal of Molecular Evolution* over the past decade has done such experimental work or model building.[10] The overwhelming percentage of papers are concerned with sequence analysis. Again, I hasten to say that sequence analysis is interesting and can tell one many things, but sequence analysis alone cannot say how complex molecular machines could have been produced in a Darwinian fashion.

If one looks at other journals, at the *Proceedings of the National Academy of Sciences*, *Cell*, the *Journal of Molecular Biology*, and so on, the story is the same. There are many, many studies comparing sequences, but very few concerning the Darwinian production of complex molecular machines. The few that do consider the problems of Darwinian evolution are invariably too broad to test rigorously. But if the scientific literature—the journals—do not contain answers to the question of how Darwinian processes could produce such intricate molecular machines, then why do many scientists believe that they can produce them? Well, it is difficult to say in detail, but certainly a part of the answer to that question is that scientists are taught, as part of their scientific training, that Darwinism is true. A good illustration can be found in the excellent textbook *Biochemistry*, by Voet and Voet. In the first chapter, where the textbook is introducing students to the biochemical view of the world, there is a marvelous, full-color drawing depicting the orthodox view of how life arose and diversified. In the top third of the drawing there are illustrated a volcano, lightning flashes, little rays of sunlight, and some gases floating around—and that, students are meant to infer, is how life started. The middle third of the picture shows a stylized drawing of a DNA molecule leading out from the origin-of-life ocean and into a bacterial cell—showing us how life developed. (The bacterium is depicted with a flagellum that, in the far-off view, looks as simple as a hair.) The bottom third of the picture shows the Garden of Eden, with a number of animals produced by evolution milling about. In their midst are a man and woman in the buff (which will no doubt attract student interest). If you look closely you see that the woman is offering the man an apple. And that, students are implicitly led to believe, is how life diversified.

But if you look through the text for serious scientific answers to how any of those processes could have occurred, you will not find them. In the *Origin of Species* at a number of points Darwin appealed to the imagination of his readers. But imagination is a two-edged sword. An imaginative person might see things that other people miss. Or he might see things that are not there. An examination of the science literature seems to show that Darwinism has become stuck in the world of imagination.

Apprehending Design

My criticisms of Darwinian theory are not really new. A number of other scientists have previously noted that the biochemistry of life is really quite complex and does not seem to fit the gradualistic mechanism that Darwin proposed. Further, it has been pointed out by others that the scientific literature contains few real explanations of the molecular foundations of life. Scientists like Stuart Kauffman of the Santa Fe Institute, James Shapiro of the University of Chicago, and Lynn Margulis of the University of Massachusetts have all stated that natural selection is not a good explanation for some aspects of life.

Where I differ from those other critics is in the alternative I propose. I have written that if you look at molecular machines, such as the cilium, the flagellum, and others, they look like they were designed—purposely designed by an intelligent agent. That proposal has attracted a bit of attention. Some of my critics have pointed out that I am a Roman Catholic and imply therefore that the proposal of intelligent design is a religious idea, not a scientific one. I disagree. I think the conclusion of intelligent design in these cases is completely empirical. That is, it is based entirely on the physical evidence, along with an understanding of how we come to conclude that an object was designed. Every day of our lives we decide, consciously or not, that some things were designed, others not. How do we do that? How do we reach those conclusions?

To begin to see how we conclude that an object or system was designed, imagine that you are walking with a friend in the woods. Suddenly your friend is pulled up by the ankle by a vine and left dangling in the air. After you cut him down, you reconstruct the situation. You see that the vine was tied to a tree limb that was bent down and held by a stake in the ground. The vine was covered by leaves so that you would not notice it, and so on. From the way the parts were arranged, you would quickly conclude that this was no accident—this was a designed trap. Your conclusion is not based on religious beliefs; it is one based firmly in the physical evidence.

Let us ask a few more questions about the vine-trap. First, who designed it? After reflecting for a minute we see that we do not have enough information to answer that question. Maybe it was an enemy of yours or your friend's; maybe it was a prankster. Without more information we cannot decide who designed the trap. Nonetheless, from the interaction of the parts of the trap, we can conclude that it was indeed designed. A second question is, when was the trap designed? Again, after a minute's thought, we see that we do not yet have enough information to answer the question. Without more data, we cannot decide if the trap was designed an hour ago, a day ago, a week ago, or longer. But again, we apprehend from the interaction of the parts of the trap the fact of design itself. The bottom line is that we need additional information to answer questions such as who, what, where, when, why, and how the trap was designed. But the fact that the trap was designed is apprehended directly from observing the system.

Although we apprehend design easily and intuitively, it can also be treated in an academically rigorous manner. An excellent start has been made in treating the design problem in a philosophically and scientifically rigorous way by the philosopher and mathematician William Dembski in his monograph *The Design Inference: Eliminating Chance through Small Probabilities*.[11]

In conclusion, I would like to hearken back to the quotations with which I began this essay. In my view there is every reason, based on hard empirical observation, to conclude with Joseph Cardinal Ratzinger that "the great projects of the living creation are not the products of chance and error. . . . [They] point to a creating Reason and show us a creating Intelligence, and they do so more luminously and radiantly today than ever before."

NOTES

[1] G. Easterbrook, "Science and God: A Warming Trend?" *Science* 277 (1997): 890–93.

[2] J. Ratzinger, *In the Beginning: A Catholic Understanding of the Story of Creation and the Fall* (Grand Rapids, Mich.: Eerdmans, 1986), pp. 54–56.

[3] C. Darwin, *On the Origin of Species* (1876; reprint, New York: New York University Press, 1988), p. 151.

[4] J. Farley, *The Spontaneous Generation Controversy from Descartes to Oparin* (Baltimore: Johns Hopkins University Press, 1977), p. 73.

[5] T. M. Devlin, *Textbook of Biochemistry* (New York: Wiley-Liss, 1997), chap. 22.3.

[6] Darwin, *Origin*, p. 154

[7] M. J. Behe, *Darwin's Black Box. The Biochemical Challenge to Evolution* (New York: Free Press, 1996), p. 39.

[8] D. Voet and J. G. Voet, *Biochemistry* (New York: J. Wiley and Sons, 1995), pp. 1252–59.

[9] Ibid., pp. 1259–60.

[10] Behe, *Darwin's*, chap. 8.

[11] W. Dembski, *The Design Inference: Eliminating Chance through Small Probabilities* (Cambridge: Cambridge University Press, 1998),

APPENDICES

MICHAEL J. BEHE

ANSWERING SCIENTIFIC
CRITICISMS OF INTELLIGENT DESIGN

Introduction

In 1859 Charles Darwin published his great work *On the Origin of Species*, in which he proposed to explain how the great variety and complexity of the natural world might have been produced solely by the action of blind physical processes. His proposed mechanism was, of course, natural selection working on random variation. In a nutshell, Darwin reasoned that the members of a species whose chance variation gave them an edge in the struggle to survive would tend to survive and reproduce. If the variation could be inherited, then over time the characteristics of the species would change. And over great periods of time, perhaps great changes would occur.

It was a very elegant idea. Nonetheless, Darwin knew his proposed mechanism could not explain everything, and in the *Origin* he gave us a criterion by which to judge his theory. He wrote: "If it could be demonstrated that any complex organ existed which could not possibly have been formed by numerous, successive, slight modifications, my theory would absolutely break down."[1] Adding, however, that he could "find out no such case", Darwin of course was justifiably interested in protecting his fledgling theory from easy dismissal, and so he threw the burden of proof on opponents to "demonstrate" that something "could not possibly" have happened—which is essentially impossible to do in science. Nonetheless let us ask, what might at least *potentially* meet Darwin's criterion? What sort of organ or system seems unlikely to be formed by "numerous, successive, slight modifications"? A good place to start is with one that is *irreducibly complex*. In *Darwin's Black Box: The Biochemical Challenge to Evolution*, I defined an irreducibly complex system as: "a single system which is composed of several well-matched, interacting parts that contribute to the basic function, and where the removal of any one of the parts causes the system to effectively cease functioning."[2]

A good illustration of an irreducibly complex system from our everyday world is a simple mechanical mousetrap. A common mousetrap has several parts, including a wooden platform, a spring with extended ends, a hammer, holding bar, and catch. Now, if the mousetrap is missing the spring, or hammer, or platform, it does not catch mice half as well as it used to, or a quarter as well. It simply does not catch mice at all. Therefore it is irreducibly complex. It turns out that irreducibly complex systems are headaches for Darwinian theory, because they are resistant to being produced in the gradual, step-by-step manner that Darwin envisioned.

As biology has progressed with dazzling speed in the past half century, we have discovered many systems in the cell, at the very foundation of life, that, like a mousetrap, are irreducibly complex. I will mention only one example here—the bacterial flagellum. The flagellum is quite literally an outboard motor that some bacteria use to swim. It is a rotary device that, like a motorboat, turns a propeller to push against liquid, moving the bacterium forward in the process. It consists of a number of parts, including a long tail that acts as a propeller, the hook region, which attaches the propeller to the drive shaft, the motor, which uses a flow of acid from the outside of the bacterium to the inside to power the turning, a stator, which keeps the structure stationary in the plane of the membrane while the propeller turns, and bushing material to allow the drive shaft to poke up through the bacterial membrane. In the absence of the hook, or the motor, or the propeller, or the drive shaft, or most of the forty different types of proteins that genetic studies have shown to be necessary for the activity or construction of the flagellum, one does not get a flagellum that spins half as fast as it used to, or a quarter as fast. Either the flagellum does not work, or it does not even get constructed in the cell. Like a mousetrap, the flagellum is irreducibly complex. And again like the mousetrap, its evolutionary development by "numerous, successive, slight modifications" is quite difficult to envision. In fact, if one examines the scientific literature, one quickly sees that no one has ever proposed a serious, detailed model for how the flagellum might have arisen in a Darwinian manner, let alone conducted experiments to test such a model. Thus in a flagellum we seem to have a serious candidate to meet Darwin's criterion. We have a system that seems very unlikely to have been produced by "numerous, successive, slight modifications".

Is there an alternative explanation for the origin of the flagellum? I think there is, and it is really not difficult to see. But in order to see it, we have to do something a bit unusual: we have to break a rule. The rule is rarely stated explicitly. But it was set forth candidly by Christian de Duve in his important 1995 book, *Vital Dust*. He wrote: "A warning: All through this book, I have tried to conform to the overriding rule that life be treated as a natural process, its origin, evolution, and manifestations, up to and including the human species, as governed by the same laws as nonliving processes."[3]

In science journals the rule is always obeyed, at least in letter, yet sometimes it is violated in spirit. For example, several years ago David DeRosier, professor of biology at Brandeis University, published a review article on the bacterial flagellum in which he remarked: "More so than other motors, the flagellum resembles a machine designed by a human."[4] That same year the journal *Cell* published a special issue on the topic of "Macromolecular Machines" (issue of February 6, 1998). On the cover of the journal was a painting of a stylized protein apparently in the shape of an animal, with a watch (perhaps William Paley's) in the foreground. Articles in the journal had titles such as "The Cell as a Collection of Protein Machines";

"Polymerases and the Replisome: Machines within Machines"; and "Mechanical Devices of the Spliceosome: Motors, Clocks, Springs and Things". By way of introduction, on the contents page was written: "Like the machines invented by humans to deal efficiently with the macroscopic world, protein assemblies contain highly coordinated moving parts."

Well, if the flagellum and other biochemical systems strike scientists as looking like "machines" that were "designed by a human" or "invented by humans", then why do we not actively entertain the idea that perhaps they were indeed designed by an intelligent being? We do not do that, of course, because it would violate the rule. But sometimes, when a fellow is feeling frisky, he throws caution to the wind and breaks a few rules. In fact, that is just what I did in *Darwin's Black Box*: I proposed that, rather than Darwinian evolution, a more compelling explanation for the irreducibly complex molecular machines discovered in the cell is that they were indeed designed, as David DeRosier and the editors of *Cell* apprehended —purposely designed by an intelligent agent. In the interests of time I will not discuss here how we apprehend design; I will just recommend to you William Dembski's book *The Design Inference*.[5]

Although I think that intelligent design is a rather obvious hypothesis, nonetheless my book seems to have caught a number of people by surprise, and so it has been reviewed rather widely. The *New York Times*, the *Washington Post*, the *Allentown Morning Call*—all the major media have taken a look at it. Unexpectedly, not everyone agreed with me. In fact, in response to my argument, several scientists have pointed to experimental results that, they maintain, either cast much doubt over the claim of intelligent design or outright falsify it. In the remainder of this paper I will discuss these counterexamples. I will show not only that they fail to support Darwinism but that they actually fit much better with a theory of intelligent design. After that, I will discuss the issue of falsifiability.

An "Evolved" Operon

Kenneth Miller, a professor of cell biology at Brown University, has recently written a book entitled *Finding Darwin's God*, in which he defends Darwinism from a variety of critics, including myself. In a chapter devoted to rebutting *Darwin's Black Box*, he correctly states that "a true acid test" of the ability of Darwinism to deal with irreducible complexity would be to "[use] the tools of molecular genetics to wipe out an existing multipart system and then see if evolution can come to the rescue with a system to replace it".[6] He then cites the careful work over the past twenty-five years of Barry Hall of the University of Rochester on the experimental evolution of a lactose-utilizing system in *E. coli*.

Here is a brief description of how the system, called the *lac* operon, functions. The *lac* operon of *E. coli* contains genes coding for several proteins that are involved in the metabolism of a type of sugar called lactose. One protein of the *lac* operon, called a permease, imports lactose through the otherwise impermeable cell membrane. Another protein is an enzyme called galactosidase, which can break down lactose to its two constituent monosaccharides, galactose and glucose, which the cell can then process further. Because lactose is rarely available in the environment, the bacterial cell switches off the genes until lactose is available. The switch is controlled by another protein called a repressor, whose gene is located next to the operon. Ordinarily the repressor binds to the *lac* operon, shutting it off by physically interfering with the operon. However, in the presence of the natural "inducer" allolactose or the artificial chemical inducer IPTG, the repressor binds to the inducer and releases the operon, allowing the *lac* operon enzymes to be synthesized by the cell.

After giving his interpretation of Barry Hall's experiments, Kenneth Miller excitedly remarks:

Think for a moment—if we were to happen upon the inter-locking biochemical complexity of the reevolved lactose system, wouldn't we be impressed by the intelligence of its design? Lac-tose triggers a regulatory sequence that switches on the synthesis of an enzyme that then metabolizes lactose itself. The products of that successful lactose metabolism then activate the gene for the lac permease, which ensures a steady supply of lactose en-tering the cell. Irreducible complexity. What good would the permease be without the galactosidase? . . . No good, of course.

By the very same logic applied by Michael Behe to other systems, therefore, we could conclude that the system had been designed. Except we *know* that it was *not* designed. We know it evolved because we watched it happen right in the laboratory! No doubt about it—the evolution of biochemical systems, even complex multipart ones, is explicable in terms of evolution. Behe is wrong.[7]

The picture Miller paints is grossly and misleadingly exagger-ated. In fact, far from being a difficulty for design, the same work that Miller points to as an example of Darwinian prowess I would cite as showing the limits of Darwinism and the need for design.

So what did Barry Hall actually do? To study bacterial evolu-tion in the laboratory, in the mid 1970s Hall produced a strain of *E. coli* in which the gene for just the galactosidase of the *lac* operon was deleted. He later wrote:

All of the other functions for lactose metabolism, including lac-tose permease and the pathways for metabolism of glucose and galactose, the products of lactose hydrolysis, remain intact, thus re-acquisition of lactose utilization requires only the evolution of a new β-galactosidase function.[8]

Thus, contrary to Miller's own criterion for "a true acid test", a multipart system was not "wiped out"—only one compo-nent of a multipart system was deleted. The *lac* permease and repressor remained intact. What is more, as we shall see, the artificial inducer IPTG was added to the bacterial culture, and an alternate, cryptic galactosidase was left intact.

Without galactosidase, Hall's cells could not grow when cultured on a medium containing only lactose as a food source. However, when grown on a plate that also included alternative nutrients, bacterial colonies could be established. When the other nutrients were exhausted, the colonies stopped growing. However, Hall noticed that after several days to several weeks, hyphae grew on some of the colonies. Upon isolating cells from the hyphae, Hall saw that they frequently had two mutations, one of which was in a gene for a protein he called "evolved β-galactosidase" ("*ebg*"), which allowed it to metabolize lactose efficiently. The *ebg* gene is located in another operon, distant from the *lac* operon, and is under the control of its own repressor protein. The second mutation Hall found was always in the gene for the *ebg* repressor protein, which caused the repressor to bind lactose with sufficient strength to de-repress the *ebg* operon.

The fact that there were two separate mutations in different genes—neither of which by itself allowed cell growth[9]—startled Hall, who knew that the odds against the mutations appearing randomly and independently were prohibitive.[10] Hall's results and similar results from other laboratories led to research in the area dubbed "adaptive mutations".[11] As Hall later wrote:

Adaptive mutations are mutations that occur in nondividing or slowly dividing cells during prolonged nonlethal selection, and that appear to be specific to the challenge of the selection in the sense that the only mutations that arise are those that provide a growth advantage to the cell. The issue of the specificity has been controversial because it violates our most basic assumptions about the randomness of mutations with respect to their effect on the cell.[12]

The mechanism(s) of adaptive mutation are currently unknown. While they are being sorted out, it seems disingenuous at best to cite results of processes which "violate our most basic assumptions about the randomness of mutations" to argue for Darwinian evolution, as Miller does.

The nature of adaptive mutation aside, a strong reason to consider Barry Hall's results to be quite modest is that the *ebg* proteins—both the repressor and galactosidase—are homologous to the *E. coli lac* proteins and overlap the proteins in activity. Both of the unmutated *ebg* proteins already bind lactose. Binding of lactose even to the unmutated *ebg* repressor induces a 100-fold increase in synthesis of the *ebg* operon.[13] Even the unmutated *ebg* galactosidase can hydrolyze lactose at a level of about 10 percent that of a "Class II" mutant galactosidase that supports cell growth.[14] These activities are not sufficient to permit growth of *E. coli* on lactose, but they already are present. The mutations reported by Hall simply enhance preexisting activities of the proteins. In a recent paper[15] Professor Hall pointed out that both the *lac* and *ebg* galactosidase enzymes are part of a family of highly conserved galactosidases, identical at thirteen of fifteen active site amino acid residues, which apparently diverged by gene duplication more than two billion years ago. The two mutations in *ebg* galactosidase that increase its ability to hydrolyze lactose change two nonidentical residues back to those of other galactosidases, so that their active sites are identical. Thus—before any experiments were done—the *ebg* active site was already a near duplicate of other galactosidases and only became more active by becoming a complete duplicate. Significantly, by phylogenetic analysis Hall concluded that those two mutations are the *only* ones in *E. coli* that confer the ability to hydrolyze lactose—that is, no other protein, no other mutation in *E. coli* will work. Hall wrote:

> The phylogenetic evidence indicates that either Asp-92 and Cys/Trp-977 are the only acceptable amino acids at those positions, or that all of the single base substitutions that might be on the pathway to other amino acid replacements at those sites are so deleterious that they constitute a deep selective valley that has not been traversed in the 2 billion years since those proteins diverged from a common ancestor.[16]

Such results hardly support extravagant claims for the creativeness of Darwinian processes.

Another critical caveat not mentioned by Kenneth Miller is that the mutants that were initially isolated would be unable to use lactose in the wild—they required the artificial inducer IPTG to be present in the growth medium. As Barry Hall states clearly,[17] in the absence of IPTG, no viable mutants are seen. The reason is that a permease is required to bring lactose into the cell. However, *ebg* only has a galactosidase activity, not a permease activity, so the experimental system had to rely on the preexisting *lac* permease. Since the *lac* operon is repressed in the absence of either allolactose or IPTG, Hall decided to include the artificial inducer in all media up to this point so that the cells could grow. Thus *the system was being artificially supported by intelligent intervention.*

The prose in Miller's book obscures the facts that most of the lactose system was already in place when the experiments began, that the system was carried through nonviable states by inclusion of IPTG, and that the system will not function without preexisting components. From a sceptical perspective, the admirably careful work of Barry Hall involved a series of micromutations stitched together by intelligent intervention. He showed that the activity of a deleted enzyme could be replaced only by mutations to a second, homologous protein with a nearly identical active site; and only if the second repressor already bound lactose; and only if the system were also artificially induced by IPTG; and only if the system were also allowed to use a preexisting permease. In my view, such results are entirely in line with the expectations of irreducible complexity requiring intelligent intervention and of limited capabilities for Darwinian processes.

Blood Clotting

A second putative counterexample to intelligent design concerns the blood clotting system. Blood clotting is a very intricate biochemical process, requiring many protein parts. I had devoted a chapter of *Darwin's Black Box* to the blood clotting cascade, claiming that it is irreducibly complex and so does not fit well within a Darwinian framework. However, Russell Doolittle, a prominent biochemist, member of the National Academy of Sciences, and expert on blood clotting, disagreed. While discussing the similarity of the proteins of the blood clotting cascade to each other in an essay in the *Boston Review* in 1997, he remarked that "the genes for new proteins come from the genes for old ones by gene duplication."[18] Doolittle's invocation of gene duplication has been repeated by many scientists reviewing my book, but it reflects a common confusion. Genes with similar sequences only suggest common descent— they do not speak to the mechanism of evolution. This point is critical to my argument and bears emphasis: *evidence of common descent is not evidence of natural selection*. Similarities among either organisms or proteins are the evidence for descent with modification, that is, for evolution. Natural selection, however, is a proposed explanation for how evolution might take place—its mechanism—and so it must be supported by other evidence if the question is not to be begged.

Doolittle then cited a paper entitled "Loss of Fibrinogen Rescues Mice from the Pleiotropic Effects of Plasminogen Deficiency".[19] (By way of brief explanation, fibrinogen is the precursor of the clot material; plasminogen is a protein that degrades blood clots.) He commented:

> Recently the gene for plaminogen [*sic*] was knocked out of mice, and, predictably, those mice had thrombotic complications because fibrin clots could not be cleared away. Not long after that, the same workers knocked out the gene for fibrinogen in another line of mice. Again, predictably, these mice were ailing, although in this case hemorrhage was the problem. And what do you think happened when these two lines of mice were crossed? For all practical purposes, the mice lacking both genes were normal! Contrary to claims about irreducible complexity, the entire ensemble of proteins is not needed. Music and harmony can arise from a smaller orchestra.[20]

The implied argument seems to be that a simpler clotting cascade might be missing factors such as plasminogen and fibrinogen, and perhaps it could be expanded into the modern clotting system by gene duplication. However, that interpretation does not stand up to a careful reading of Bugge et al.[21]

In their paper Bugge et al. note that the lack of plasminogen in mice results in many problems, such as high mortality, ulcers, severe thrombosis, and delayed wound healing. On the other hand, lack of fibrinogen results in failure to clot, frequent hemorrhage, and death of females during pregnancy. The point of Bugge et al. was that if one crosses the two knockout strains, producing plasminogen-plus-fibrinogen deficiency in individual mice, the mice do not suffer the many problems that afflict mice lacking plasminogen alone. Since the title of the paper emphasized that mice are "rescued" from some ill effects, one might be misled into thinking that the double-knockout mice were normal. They are not. As Bugge et al. state in their abstract, "Mice deficient in plasminogen and fibrinogen are phenotypically indistinguishable from fibrinogen-deficient mice."[22] In other words, the double-knockouts have all the problems that mice lacking only fibrinogen have: they do not form clots, they hemorrhage, and the females die if they become pregnant.[23] They are definitely not promising evolutionary intermediates.

The probable explanation is straightforward. The pathological symptoms of mice missing just plasminogen apparently are caused by uncleared clots. But fibrinogen-deficient mice cannot form clots in the first place. So problems due to uncleared clots do not arise either in fibrinogen-deficient mice or in mice that lack both plasminogen and fibrinogen. Nonetheless, the severe problems that attend lack of clotting in fibrinogen-deficient mice continue in the double-knockouts. Pregnant females still perish.

Most important for the issue of irreducible complexity, however, is that the double-knockout mice do not merely have a less sophisticated but still functional clotting system. They have no functional clotting system at all. They are not evidence for the Darwinian evolution of blood clotting. Therefore my argument, that the system is irreducibly complex, is unaffected by that example.

Other work from the same laboratory is consistent with the view that the blood-clotting cascade is irreducibly complex. Experiments with "knock-out" mice in which the genes for other clotting components, called tissue factor and prothrombin, have been deleted separately show that those components are required for clotting, and in their absence the organism suffers severely.[24]

Falsifiability

Let us now consider the issue of falsifiability. Let me say up front that I know most philosophers of science do not regard falsifiability as a necessary trait of a successful scientific theory. Nonetheless, falsifiabilty is still an important factor to consider since it is nice to know whether or not one's theory can be shown to be wrong by contact with the real world.

A frequent charge made against intelligent design is that it is unfalsifiable, or untestable. For example, in its recent booklet *Science and Creationism*, the National Academy of Sciences writes: "[I]ntelligent design . . . [is] not science because [it is] not testable by the methods of science."[25] Yet that claim seems to be at odds with the criticisms I have just summarized. Clearly, Russell Doolittle and Kenneth Miller advanced scientific arguments aimed at falsifying intelligent design. If the results of Bugge et al.[26] had been as Doolittle first thought, or if Barry Hall's work had indeed shown what Miller implied, then they correctly believed that my claims about irreducible complexity would have suffered quite a blow.

Now, one cannot have it both ways. One cannot say both that intelligent design is unfalsifiable (or untestable) and that there is evidence against it. Either it is unfalsifiable and floats serenely beyond experimental reproach, or it can be criticized on the basis of our observations and is therefore testable. The fact that critical reviewers advance scientific arguments against intelligent design (whether successfully or not) shows that intelligent design is indeed falsifiable. What is more, it is wide open to falsification by a series of rather straightforward laboratory experiments such as those that Miller and Doolittle pointed to, which is exactly why they pointed to them.

Now let us turn the tables and ask: How could one falsify the claim that a particular biochemical system was produced by a Darwinian process? Kenneth Miller announced an "acid test" for the ability of natural selection to produce irreducible complexity. He then decided that the test was passed and unhesitatingly proclaimed intelligent design to be falsified. But if, as it certainly seems to me, *E. coli* actually fails the lactose-system "acid test", would Miller consider Darwinism to be falsified? Almost certainly not. He would surely say that Barry Hall started with the wrong bacterial species or used the wrong selective pressure, and so on. So it turns out that his "acid test" was not a test of Darwinism; it tested only intelligent design.

The same one-way testing was employed by Russell Doolittle. He pointed to the results of Bugge et al. to argue against intelligent design. But when the results turned out to be the opposite of what he had originally thought, Professor Doolittle did not abandon Darwinism.

It seems then, perhaps counterintuitively to some, that intelligent design is quite susceptible to falsification, at least on the points under discussion. Darwinism, on the other hand, seems quite impervious to falsification. The reason for that can be seen when we examine the basic claims of the two ideas with regard to a particular biochemical system like, say, the bacterial flagellum. The claim of intelligent design is that "*No* unintelligent process could produce this system." The claim of Darwinism is that "*Some* unintelligent process could produce this system." To falsify the first claim, one need only show that at least one unintelligent process could produce the system. To falsify the second claim, one would have to show the system could not have been formed by any of a potentially infinite number of possible unintelligent processes, which is effectively impossible to do.

The danger of accepting an effectively unfalsifiable hypothesis is that science has no way to determine if the belief corresponds to reality. In the history of science, the scientific community has believed in any number of things that were in fact not true, not real, for example, the universal ether. If there were no way to test those beliefs, the progress of science might be substantially and negatively affected. If, in the present case, the expansive claims of Darwinism are in reality not true, then its unfalsifiability will cause science to bog down, as I believe it has.

So, what can be done? I do not think that the answer is never to investigate a theory that is unfalsifiable. After all, although it is unfalsifiable, Darwinism's claims are potentially positively demonstrable. For example, if some scientist conducted an experiment showing the production of a flagellum (or some equally complex system) by Darwinian processes, then the Darwinian claim would be affirmed. The question only arises in the face of negative results.

I think several steps can be prescribed. First of all, one has to be aware—raise one's consciousness—about when a theory is unfalsifiable. Second, as far as possible, an advocate of an unfalsifiable theory should try as diligently as possible to demonstrate positively the claims of the hypothesis. Third, one needs to relax Darwin's criterion from this:

> If it could be demonstrated that any complex organ existed which could not possibly have been formed by numerous, successive, slight modifications, my theory would absolutely break down.

to something like this:

> If a complex organ exists which seems *very unlikely* to have been produced by numerous, successive, slight modifications, and if no experiments have shown that it or comparable structures can

be so produced, then maybe we are *barking up the wrong tree*. So,
LET'S BREAK SOME RULES!

Of course, people will differ on the point at which they decide to break rules. But at least with the realistic criterion there
could be evidence against the unfalsifiable. At least then people
like Doolittle and Miller would run a risk when they cite an
experiment that shows the opposite of what they had thought.
At least then science would have a way to escape from the rut
of unfalsifiability and think new thoughts.

NOTES

[1] C. Darwin, *The Origin of Species* (1859; reprint, New York, Bantam Books), p. 154.

[2] M. J. Behe, *Darwin's Black Box: The Biochemical Challenge to Evolution* (New York: Free Press, 1996), p. 39.

[3] C. de Duve, *Vital Dust: Life as a Cosmic Imperative* (New York: Basic Books: 1995), p. xiv.

[4] D. J. DeRosier, "The Turn of the Screw: The Bacterial Flagellar Motor", *Cell* 93 (1998): 17–20.

[5] W. A. Dembski, *The Design Inference: Eliminating Chance through Small Probabilities* (Cambridge: Cambridge University Press, 1998).

[6] K. R. Miller, *Finding Darwin's God: A Scientist's Search for Common Ground between God and Evolution* (New York: Cliff Street Books, 1999), p. 145.

[7] Ibid., pp. 146–47.

[8] B. G. Hall, "Experimental Evolution of Ebg Enzyme Provides Clues about the Evolution of Catalysis and to Evolutionary Potential", *FEMS Microbiology Letters* 174 (1999): 1–8.

[9] B. G. Hall, "Evolution of a Regulated Operon in the Laboratory", *Genetics* 101 (1982): 335–44.

[10] B. G. Hall, "Evolution on a Petri Dish: The Evolved β-Galactosidase System as a Model for Studying Acquisitive Evolution in the Laboratory", in *Evolutionary Biology*, ed. M. K. Hecht, B. Wallace, and G. T. Prance (New York: Plenum Press, 1982), pp. 85–150.

[11] P. L. Foster, "Mechanisms of Stationary Phase Mutation: A Decade of Adaptive Mutation", *Annual Review of Genetics* 33 (1999): 57–88.

[12] B. G. Hall, "On the Specificity of Adaptive Mutations", *Genetics* 145 (1997): 39–44.

[13] Hall, "Regulated Operon".

[14] Hall, "Experimental Evolution".

[15] Ibid.

[16] Ibid.

[17] Hall, "Petri Dish".

[18] R. F. Doolittle, "A Delicate Balance", *Boston Review*, February/March 1997, pp. 28–29.

[19] T. H. Bugge, K. W. Kombrinck, M. J. Flick, C. C. Daugherty, M. J. Danton, and J. L. Degen, "Loss of Fibrinogen Rescues Mice from the Pleiotropic Effects of Plasminogen Deficiency", *Cell* 87 (1996): 709–19.

[20] Doolittle, "Delicate Balance".

[21] Bugge et al., "Loss".

[22] Ibid.

[23] T. T. Suh, K. Holmback, N. J. Jensen, C. C. Daugherty, K. Small, D. I. Simon, S. Potter, and J. L. Degen, "Resolution of Spontaneous Bleeding Events but Failure of Pregnancy in Fibrinogen-Deficient Mice", *Genes and Development* 9 (1995): 2020–33.

[24] T. H. Bugge, Q. Xiao, K. W. Kombrinck, M. J. Flick, K. Holmback, M. J. Danton, M. C. Colbert, D. P. Witte, K. Fujikawa, E. W. Davie, and J. L. Degen, "Fatal Embryonic Bleeding Events in Mice Lacking Tissue Factor, the Cell-Associated Initiator of Blood Coagulation", *Proceedings of the National Academy of Sciences of the United States of America* 93 (1996): 6258–63; W. Y. Sun, D. P. Witte, J. L. Degen, M. C. Colbert, M. C. Burkart, K. Holmback, Q. Xiao, T. H. Bugge, and S. J. Degen, "Prothrombin Deficiency Results in Embryonic and Neonatal Lethality in Mice", *Proceedings of the National Academy of Sciences of the United States of America* 95 (1998): 7597–7602.

[25] National Academy of Sciences, *Science and Creationism: A View from the National Academy of Sciences* (Washington, D.C.: National Academy Press, 1999), p. 25.

[26] Bugge et al., "Loss".

Stephen C. Meyer

THE SCIENTIFIC STATUS
OF INTELLIGENT DESIGN

The Methodological Equivalence
of Naturalistic and Non-Naturalistic
Origins Theories

Throughout the *Origin of Species*, Darwin repeatedly argues against the scientific status of the received "theory of Creation". He often faults his creationist rivals, not just for their inability to devise explanations for certain biological data, but for their inability to offer *scientific* explanations at all. Indeed, some of Darwin's arguments for descent with modification depended, not on newly discovered facts unknown to the special creationists, but upon facts such as fossil progression, homology, and biogeographical distribution that had neither stymied nor puzzled many creationists but that, in Darwin's view, creationists could not explain in a properly scientific way.[1] What Darwin questioned in his attack against creationism was not just, to put the issue in modern terms, the "empirical adequacy" of then current creationist theories, but rather the methodological (and therefore scientific) legitimacy of the creationist program itself. Thus, Darwin would emphatically dismiss the

For helpful comments and criticisms I would like to thank Ed Olson, Forrest Baird, Dale Bruner, Bill Dembski, Norman Krebbs, J. P. Moreland, Paul Nelson, and Jitse van der Meer. For assistance with typing references I would like to thank Lorrie Nelson. For generous research support I would like to thank the Pascal Centre in Ontario, Canada, and C. Davis Weyerhaeuser.

creationist account of homology, for example, by saying "but that is not a *scientific* explanation."[2]

Underlying Darwin's repudiation of creationist legitimacy lay an entirely different conception of science than had prevailed among earlier naturalists.[3] Darwin's attacks on his creationist and idealist opponents in part expressed and in part established an emerging positivistic[4] "episteme" in which the mere mention of unverifiable "acts of divine will" or "the plan of creation" would increasingly serve to disqualify theories from consideration as science qua science. This decoupling of theology from science and the redefinition of science that underlay it was justified less by argument than by an implicit assumption about the characteristic features of all scientific theories—features that presumably could distinguish theories of a properly scientific (that is, positivistic) bent from those tied to unwelcome metaphysical or theological moorings. Thus, both in the *Origin* and in subsequent letters one finds Darwin invoking a number of ideas about what constitutes a properly scientific explanation in order to characterize creationist theories as inherently "unscientific". For Darwin the *in-principle* illegitimacy of creationism was demonstrated by perceived deficiencies in its method of inquiry, such as its failure to explain by reference to natural law[5] and its postulation of unobservable causes and explanatory entities such as mind, purpose, or "the plan of creation".[6]

Future defenders of Darwinism would expand this strategy.[7] Throughout the twentieth century those attempting to defend naturalistic evolutionary theories from challenge by *any* non-naturalistic origins theory have often invoked various norms of scientific practice. These norms have typically been derived from the philosophy of science, most particularly from the logical positivists or the neopositivists (such as Sir Karl Popper or Carl Hempel). Both the positivistic standard of verifiability and the neopositivistic standards of falsifiability and lawlike explanation have functioned as methodological yardsticks or "demarcation criteria" for measuring, and finding deficient, all theories of creation or even theories of intelligent design. Such theories have been declared "unscientific by definition" on numerous philosophical and methodological grounds.

The use by evolutionary biologists of so-called demarcation arguments—that is, arguments that purport to distinguish science from pseudoscience, metaphysics, or religion—is both ironic and problematic from the point of view of the philosophy of science. It is ironic because many of the demarcation criteria that have been used against nonnaturalistic theories of origin can be deployed with equal warrant against strictly naturalistic evolutionary theories. Indeed, a corpus of literature now exists devoted to assessing whether neo-Darwinism, with its distinctively probabilistic and historical dimensions, is scientific when measured against various conceptions of science.[8] Some have wondered whether the use of narrative explanation in evolutionary biology constitutes a departure from a strict reliance upon natural law. Others have asked whether neo-Darwinism is falsifiable, or whether it makes true or risky predictions. In 1974, Sir Karl Popper declared neo-Darwinian evolutionary theory "untestable" and classified it as a "metaphysical research programme". While he later revised his judgment, he did so only after liberalizing his notion of falsifiability to allow the weaker notion of "falsifiability in principle" to count as a token of scientific status.

The use of demarcation arguments to settle the origins controversy is also problematic because the whole enterprise of demarcation has now fallen into disrepute. Attempts to locate methodological "invariants" that provide a set of necessary and sufficient conditions for distinguishing true science from pseudoscience have failed.[9] Most philosophers of science now recognize that neither verifiability nor testability (nor falsifiability) nor the use of lawlike explanation (nor any other criterion) can suffice to define scientific practice.[10]

Nevertheless, philosophical arguments about what does or does not constitute science continue to play a vital role in persuading biologists that alternative scientific explanations do not and (in the case of nonnaturalistic or nonmaterialistic explanations) *can* not exist for the origin of biological form and structure. Indeed, demarcation criteria continue to be cited by modern biologists as reasons for disregarding the possibility of intelligent design as a theory of biological origins.[11]

This essay will examine the in-principle case against the scientific status of intelligent design. It will examine several of the methodological criteria that have been advanced as means of distinguishing the scientific status of naturalistic evolutionary theories from nonnaturalistic theories such as intelligent design, special creation, progressive creation, and theistic evolution. I will argue that attempts to make distinctions of scientific status a priori on methodological grounds inevitably fail and, instead, that a general equivalence of method exists between these two broadly competing approaches to origins. In so doing, I will attempt to shed light on the specific question of whether a scientific theory of intelligent design could be formulated, or whether methodological objections, forever and in principle, render this possibility "self-contradictory nonsense", as Ruse, Stent, Gould, and others have claimed (of, at least, scientific creationism).[12] Throughout this paper, I will use the alliterative terms "design" and "descent" as a convenient shorthand to distinguish (a) theories that invoke the efficient causal action of an intelligent agent (whether divine or otherwise) as part of the explanation for the origin of biological form and complexity from (b) theories (such as Darwin's "descent with modification") that rely *exclusively* on naturalistic processes to explain the origin of biological form and complexity.[13]

By way of qualification, it should be noted that by defending the methodological and scientific legitimacy of intelligent design, this essay is not seeking to rehabilitate the empirically inadequate biology of many nineteenth-century creationists or their belief in the absolute fixity of species; nor is it attempting to endorse modern young-earth geology. The following analysis concerns the methodological legitimacy of "design" in principle as defined above, not the empirical adequacy of specific theories that might invoke intelligent design in the process of making other empirical claims.

The methodological equivalence of intelligent design and naturalistic descent will be suggested in three stages by three lines of argument. First, the reasons for the failure of demarcation arguments within philosophy of science generally will be examined and recapitulated. This analysis will suggest that attempts to distinguish the scientific status of design and descent a priori may well be suspect from the outset on philosophical grounds. Second, an examination of specific demarcation arguments that have been employed against design will follow. It will be argued that not only do these arguments fail, but they do so in such a way as to suggest an equivalence between design and descent with respect to several features of allegedly proper scientific practice —that is, intelligent design and naturalistic descent will be shown equally capable or incapable of meeting different demarcation standards, provided such standards are applied disinterestedly. Third, design and descent will be compared in light of recent work on the logical and methodological character of historical inquiry. This analysis will show that the mode of inquiry utilized by advocates of both design and descent conforms closely to that evident in many other characteristically historical disciplines. Thus a more fundamental methodological equivalence between design and descent will emerge as a result of methodological analysis of the historical sciences.

Part 1: The General Failure of Demarcation Arguments

To show that design "can never be considered a scientific pursuit,"[14] biologists and others have asserted that design does not meet certain objective criteria of scientific method or practice. In short, biologists have employed so-called demarcation arguments to separate a scientific approach to origins (descent) from an allegedly nonscientific approach (design). While an examination of the particular criteria employed in such arguments will not concern us in the first part of this essay, the general practice of demarcation will.

From the standpoint of the philosophy of science, the use of demarcation arguments is generally problematic. Historically, attempts to find methodological "invariants" that provide a set of necessary and sufficient conditions for distinguishing true science from pseudoscience have failed.[15] Moreover, most current demarcation arguments presuppose an understanding of how science operates that reflects the influence of a philosophy of science known as logical positivism. Yet since the 1950s philosophers of science have decisively rejected positivism for a number of very good reasons (see below). As a result, the enterprise of demarcation has generally fallen into disrepute among philosophers of science.

In his essay "The Demise of the Demarcation Problem", philosopher of science Larry Laudan gives a brief but thorough sketch of the different grounds that have been advanced during the history of science for distinguishing science from nonscience.[16] He notes that the first such grounds concerned the degree of certainty associated with scientific knowledge. Science, it was thought, could be distinguished from nonscience because science produced certainty whereas other types of inquiry such as philosophy produced opinion. Yet this approach to demarcation ran into difficulties as scientists and philosophers gradually realized the fallible nature of scientific disciplines and theories. Unlike mathematicians, scientists rarely provide strict logical demonstrations (deductive proofs) to justify their theories. Instead, scientific arguments often utilize inductive inference and predictive testing, neither of which produces certainty. As Owen Gingerich has argued, much of the reason for Galileo's conflict with the Vatican stemmed from Galileo's inability to meet scholastic standards of deductive certainty—a standard that he regarded as neither relevant to nor attainable by scientific reasoning.[17] Similar episodes subsequently made it clear that science does not necessarily possess a superior epistemic status; scientific knowledge, like other knowledge, is subject to uncertainty.

By the nineteenth century, attempts to distinguish science from nonscience had changed. No longer did demarcationists attempt to characterize science on the basis of the superior epistemic status of scientific theories; rather, they attempted to do so on the basis of the superior methods science employed to produce theories. Thus science came to be defined by reference to its method, not its content. Demarcation criteria became methodological rather than epistemological.[18]

Nevertheless, this approach also encountered difficulties, not the least of which was a widespread disagreement about what the method of science really is. If scientists and philosophers cannot agree about what *the* scientific method is, how can they disqualify disciplines that fail to use it? Moreover, as the discussion of the historical sciences in part 3 of this essay will make clear, there may well be more than one scientific method. If that is so, then attempts to mark off science from nonscience using a single set of methodological criteria will most likely fail. The existence of a variety of scientific methods raises the possibility that no single methodological characterization of science may suffice to capture the diversity of scientific practice. Using a single set of methodological criteria to assess scientific status could therefore result in the disqualification of some disciplines already considered to be scientific.[19]

As problems with using methodological considerations grew, demarcationists shifted their focus again. Beginning in the 1920s, philosophy of science took a linguistic or semantic turn. The logical positivist tradition held that scientific theories could be distinguished from nonscientific theories, not because scientific theories had been produced via unique or superior methods, but because such theories were more meaningful. Logical positivists asserted that all meaningful statements are either empirically verifiable or logically undeniable. According to this "verificationist criterion of meaning", scientific theories are more meaningful than philosophical or religious ideas, for example, because scientific theories refer to observable entities such as planets, minerals, and birds, whereas philosophy and religion refer to such unobservable entities as God, truth, and morality.

Yet as is now well known, positivism soon self-destructed. Philosophers came to realize that positivism's verificationist criterion of meaning did not achieve its own standard. That is, the assumptions of positivism turn out to be neither empirically verifiable nor logically undeniable. Furthermore, positivism's verificationist ideal misrepresented much actual scientific practice. Many scientific theories refer to unverifiable and unobservable entities such as forces, fields, molecules, quarks, and universal laws. Meanwhile, many disreputable theories (for example, the flat-earth theory) appeal explicitly to "common-sense" observations. Clearly, positivism's verifiability criterion would not achieve the demarcation desired.

With the death of positivism in the 1950s, demarcationists took a different tack. Other semantic criteria emerged, such as Sir Karl Popper's falsifiability. According to Popper, scientific theories were more meaningful than nonscientific ideas because they referred only to empirically falsifiable entities.[20] Yet this, too, proved to be a problematic criterion. First, falsification turns out to be difficult to achieve. Rarely are the core commitments of theories directly tested via prediction. Instead, predictions occur when core theoretical commitments are conjoined with auxiliary hypotheses, thus always leaving open the possibility that auxiliary hypotheses, not core commitments, are responsible for failed predictions.

Newtonian mechanics, for example, assumed as its core three laws of motion and the theory of universal gravitation. On the basis of these, Newton made a number of predictions about the positions of planets in the solar system. When observations failed to corroborate some of his predictions, he did not reject his core assumptions. Instead, he scrutinized some of his auxiliary hypotheses to explain the discrepancies between theory and observation. For example, he examined his working assumption that planets were perfectly spherical and influenced only by gravitational force. As Imre Lakatos has shown, Newton's refusal to repudiate his core in the face of anomalies enabled him to refine his theory and eventually led to its tremendous success.[21] Newton's refusal to accept putatively falsifying results certainly did not call into question the scientific status of his gravitational theory or his three laws.

The function of auxiliary hypotheses in scientific testing suggests that many scientific theories, including those in so-called hard sciences, may be very difficult, if not impossible, to falsify conclusively. Yet many theories that have been falsified in practice via the consensus judgment of the scientific community must qualify as scientific according to the falsifiability criterion. Since they have been falsified, they are obviously falsifiable, and since they are falsifiable, they would seem to be scientific.[22]

And so it has gone generally with demarcation criteria. Many theories that have been repudiated on evidential grounds express the very epistemic and methodological virtues (testability, falsifiability, observability, and so on) that have been alleged to characterize true science. Many theories that are held in high esteem lack some of the allegedly necessary and sufficient features of proper science. As a result,[23] with few exceptions[24] most contemporary philosophers of science regard the question "What methods distinguish science from nonscience?" as both intractable and uninteresting. What, after all, is in a name? Certainly not automatic epistemic warrant or authority. Thus philosophers of science have increasingly realized that the real issue is not whether a theory is scientific but whether it is true or warranted by the evidence. Thus, as Martin Eger has summarized, "Demarcation arguments have collapsed. Philosophers of science don't hold them anymore. They may still enjoy acceptance in the popular world, but that's a different world."[25]

The "demise of the demarcation problem", as Laudan calls it, implies that the use of positivistic demarcationist arguments by evolutionists is, at least prima facie, on very slippery ground. Laudan's analysis suggests that such arguments are not likely to succeed in distinguishing the scientific status of descent vis-à-vis design or anything else for that matter. As Laudan puts it, "If we could stand up on the side of reason, we ought to drop terms like 'pseudo-science.' . . . They do only emotive work for us."[26]

If philosophers of science such as Laudan are correct, a stalemate exists in our analysis of design and descent. Neither can automatically qualify as science; neither can be necessarily disqualified either. The a priori methodological merit of design and descent are indistinguishable if no agreed criteria exist by which to judge their merits.

Yet lacking any definite metric, one cannot yet say that design and descent are methodologically equivalent in any non-trivial sense. In order to make this claim we must compare design and descent against some specific standards. Let us now consider the specific demarcation arguments that have been erected against design. For though demarcation arguments have been discredited by philosophers of science generally, they still enjoy wide currency in the scientific and "popular world",[27] as the following section will make abundantly clear.

Part 2: Specific Demarcation
Arguments against Design

Despite the consensus among philosophers of science that the demarcation problem is both intractable and ill-conceived, many scientists continue to invoke demarcation criteria to discredit quacks, cranks, and those otherwise perceived as intellectual opponents. Yet to the average working scientist Laudan's arguments against demarcation may seem counterintuitive at best. On the surface it may appear that there ought to be some unambiguous criteria for distinguishing such dubious pursuits as parapsychology, astrology, and phrenology from established sciences such as physics, chemistry, and astronomy. That most philosophers of science say that there are not such criteria only confirms the suspicions many scientists have about philosophers of science. After all, do not some philosophers of science say that scientific truth is determined by social and cultural context? Do not some even deny that science describes an objective reality?

Well, as it turns out, one does not need to adopt a relativistic or antirealist view of science to accept what Laudan and others say about the demarcation problem. Indeed, the two positions are logically unrelated. Laudan is not arguing that all scientific theories have equal warrant (quite the reverse) or that scientific theories never refer to real entities. Instead, he simply says that one cannot define science in such a way as to confer automatic epistemic authority on favored theories simply because they happen to manifest features alleged to characterize all "true science". When evaluating the warrant or truth claims of theories, we cannot substitute abstractions about the nature of science for empirical evaluation.

Nevertheless, establishing Laudan's general thesis is not the main purpose of this essay. This essay is not seeking to establish the impossibility of demarcation in general but the methodological equivalence of intelligent design and naturalistic descent. Since some may yet doubt that demarcation *always* fails, the following section will examine some of the specific demarcation arguments that have been deployed against design by proponents of descent.[28] It will suggest that these arguments fail to provide any grounds for distinguishing the methodological merit of one over the other and, instead, that careful analysis of these arguments actually exposes reasons for regarding design and descent as methodologically equivalent. Indeed, the following analysis will suggest that metaphysically neutral criteria do not exist that can define science narrowly enough to disqualify theories of design *tout court* without also disqualifying theories of descent on identical grounds.

Unfortunately, to establish this conclusively would require

an examination of all the demarcation arguments that have been used against design. And indeed, an examination of evolutionary polemic reveals many such arguments. Design or creationist theories have been alleged to be necessarily unscientific because they (a) do not explain by reference to natural law,[29] (b) invoke unobservables,[30] (c) are not testable,[31] (d) do not make predictions,[32] (e) are not falsifiable,[33] (f) provide no mechanisms,[34] (g) are not tentative,[35] and (h) have no problem-solving capability.[36]

Due to space constraints, a detailed analysis of only the first three arguments will be possible. Nevertheless, an extensive analysis of (a), (b), and (c) will follow. These three have been chosen because each can be found in one form or another all the way back to the *Origin of Species*. The first one, (a), is especially important because the others derive from it—a point emphasized by Michael Ruse,[37] perhaps the world's most ardent evolutionary demarcationist. Consequently an analysis of assertion (a) will occupy the largest portion of this section.[38] There will also be a short discussion of arguments (d), (e), and (f) and references to literature refuting (g) and (h). Thus, while an exhaustive analysis of all demarcationist arguments will not be possible here, enough will be said to allow us to conclude that the principal arguments employed against design do not succeed in impugning its scientific status without either begging the question or undermining the status of descent as well.

Explanation via natural law. Now let us examine the first, and according to Michael Ruse[39] most fundamental, of the arguments against the possibility of a scientific theory of design. This argument states: "Scientific theories must explain by natural law. Because design or creationist theories do not do so, they are necessarily unscientific."

This argument invokes one of the principal criteria of science adopted by Judge William Overton after hearing the testimony of philosopher of science Michael Ruse in the Arkansas creation-science trial of 1981–1982.[40] As late as March 1992,

Ruse continued to assert "must explain via natural law" as a demarcation criterion, despite criticism from other philosophers of science such as Philip Quinn and Larry Laudan.[41] Ruse has argued that to adopt the scientific outlook, one must accept that the universe is subject to natural law and, further, that one must never appeal to an intervening agency as an explanation for events. Instead, one must always look to what he calls "unbroken law" if one wishes to explain things in a scientific manner.

There are several problems with this assertion and the conception of science that Ruse assumes.[42] In particular, Ruse seemed to assume a view of science that equates scientific laws with explanations. There are two problems with this view and correspondingly two main reasons that "explains via natural law" will not do as a demarcation criterion.

First, many laws are descriptive and not explanatory. Many laws describe regularities but do not explain why the regular events they describe occur. A good example of this drawn from the history of science is the universal law of gravitation, which Newton himself freely admitted did not explain but instead merely described gravitational motion. As he put it in the "General Scholium" of the second edition of the *Principia*, "I do not feign hypotheses"—in other words, "I offer no explanations."[43] Insisting that science must explain by reference to "natural law" would eliminate from the domain of the properly scientific all fundamental laws of physics that describe mathematically, but do not explain, the phenomena they "cover".[44] For the demarcationist this is a highly paradoxical and undesirable result, since much of the motivation for the demarcationist program derives from a desire to ensure that disciplines claiming to be scientific match the methodological rigor of the physical sciences. While this result might alleviate the "physics envy" of many a sociologist, it does nothing for demarcationists except defeat the very purpose of their enterprise.

There is a second reason that laws cannot be equated with

explanations or causes. This, in turn, gives rise to another reason that science cannot be identified only with those disciplines that explain via natural law. Laws cannot be equated with explanations, not just because many laws do not explain, but also because many explanations of particular events, especially in applied or historical science, may not utilize laws.[45] While scientists may often use laws to assess or enhance the plausibility of explanations of particular events, analysis of the logical requirements of explanation has made clear that the citation of laws is not necessary to many such explanations.[46] Instead, many explanations of particular events or facts, especially in the historical sciences, depend primarily, even exclusively, upon the specification of past causal conditions and events rather than laws to do what might be called the "explanatory work". That is, citing past causal events often explains a particular event better than, and sometimes without reference to, a law or regularity in nature.[47]

One reason laws play little or no role in many historical explanations is that many particular events come into existence via a series of events that will not regularly reoccur. In such cases laws are not relevant to explaining the contrast between the event that has occurred and what could have or might have ordinarily been expected to occur. For example, a historical geologist seeking to explain the unusual height of the Himalayas will cite particular antecedent factors that were present in the case of the Himalayan orogeny but were absent in other mountain-building episodes. Knowing the laws of geophysics relevant to mountain building generally will aid the geologist very little in accounting for the contrast between the Himalayan and other orogenies, since such laws would presumably apply to all mountain-building episodes. What the geologist needs in the search for an explanation in this case is not knowledge of a general law but evidence of a unique or distinctive set of past conditions.[48] Thus geologists have typically explained the unique height of the Himalayas by reference to the past position of the Indian and Asian land masses (and plates) and the subsequent collision that occurred between them.

The geologist's situation is very similar to that faced by historians generally. Consider the following factors that might help explain why World War I began: the ambition of Kaiser Wilhelm's generals, the Franco-Russian defense pact, and the assassination of Archduke Ferdinand. Note that such possible explanatory factors invariably involve the citation of past events, conditions, or actions rather than laws. Invoking past events as causes in order to explain subsequent events or present evidences is common both in history and in natural scientific disciplines such as historical geology. As Michael Scriven has shown, one can often know what caused something even when one cannot relate causes and effects to each other in formal statements of law.[49] Similarly, William Alston has shown that laws alone often do not explain particular events even when we have them.[50] The law "Oxygen is necessary to combustion" does not explain why a particular building burned down at a particular place and time.[51] To explain such a particular fact requires knowing something about the situation just before the fire occurred. It does little good to know scientific laws; what one requires is information concerning, for example, the presence of an arsonist or the lack of security at the building or the absence of a sprinkler system. Thus Alston concludes that to equate a law with an explanation or cause "is to commit a 'category mistake' of the most flagrant sort".[52]

Perhaps another example will help. If one wishes to explain why astronauts were able to fly to the moon when apples usually fall to the earth, one will not primarily cite the law of gravity. Such a law is far too general to be primarily relevant to explanation in this context, because the law allows for a vast array of possible outcomes depending on initial and boundary conditions. The law stating that all matter gravitates according to an inverse square law is consistent both with an apple falling to the earth and with an astronaut flying to the moon. Explaining why the astronaut flew when apples routinely fall, therefore, requires more than citing the law, since the law is presumed operative in both situations. Accounting for the differing outcomes—the falling apple and the flying astronaut—will require references to the antecedent conditions and events that differed in the two situations. Indeed, explanation in this case involves an accounting of the way engineers have used technology to alter the *conditions* affecting the astronauts to allow them to overcome the constraints that gravity ordinarily imposes on earthbound objects.

Such examples suggest that many explanations of particular events—explanations that occur frequently in fields already regarded as scientific—such as cosmology, archaeology, historical geology, applied physics and chemistry, origin-of-life studies and evolutionary biology—would lose their scientific status if Ruse's criterion of "explains via natural law" were accepted as normative to all scientific practice.

Consider an example from evolutionary biology that impinges directly on our discussion. Stephen Jay Gould, Mark Ridley, and Michael Ruse argue that the "fact of evolution"[53] is secure even if an adequate theory has not yet been formulated to describe or explain how large-scale biological change generally occurs. Like Darwin, modern evolutionary theorists insist that the question whether evolution[54] did occur can be separated logically from the question of the means by which nature generally achieves biological transformations. Evolution in one sense—historical continuity or common descent—is asserted to be a well-established scientific theory[55] because it alone explains a diverse class of present data (fossil progression, homology, biogeographical distribution, and so on), even if biologists cannot yet explain how evolution in another sense—a general process or mechanism of change—occurs. Some have likened the logical independence of common descent and natural selection to the logical independence of continental drift and plate tectonics. In both the geological situation and the biological there exist theories about *what happened*, which explain why we observe many present facts, and separate theories that explain *how* things *could have* happened as they apparently did. Yet the former purely historical explanations do not require the latter nomological[56] or mechanistic explanations to legitimate themselves. Common descent explains some facts well, even if nothing yet explains how the transformations it requires could have occurred.

This example again illustrates why historical explanations do not require laws.[57] More important, it also demonstrates why Ruse's demarcation criterion proves fatal to the very Darwinism he is seeking to protect. Common descent, arguably a central thesis of the *Origin of Species*, does not explain by natural law. Common descent explains by postulating a hypothetical pattern of historical events that, if actual, would account for a variety of presently observed data. Darwin himself refers to common descent as the *vera causa* (that is, the actual cause or explanation) for a diverse set of biological observations.[58] In Darwin's historical argument for common descent, as with historical explanations generally, postulated past causal events (or patterns thereof) do the primary explanatory work. Laws do not.[59]

At this point the evolutionary demarcationist might grant the explanatory function of antecedent events but deny that scientific explanations can invoke *supernatural* events. To postulate naturally occurring past events is one thing, but to postulate supernatural events is another. The first leaves the laws of nature intact; the second does not and thus lies beyond the bounds of science. As Ruse and Richard Lewontin have argued, miraculous events are unscientific because they violate or contradict the laws of nature, thus making science impossible.[60]

Many contemporary philosophers disagree with Ruse and Lewontin about this, as have a number of good scientists over the years—Isaac Newton and Robert Boyle, for example. The action of agency (whether divine or human) need not violate the laws of nature; in most cases it merely changes the initial and boundary conditions on which the laws of nature operate.[61] But this issue must be set aside for the moment. For now it will suffice merely to note that the criterion of demarcation has subtly shifted. No longer does the demarcationist repudiate design as unscientific because it does not "explain via natural law"; now the demarcationist rejects intelligent design because it does not "explain naturalistically". To be scientific a theory must be naturalistic.

But why is this the case? Surely the point at issue is whether there are independent and metaphysically neutral grounds for disqualifying theories that invoke nonnaturalistic events—such as instances of agency or intelligent design. To assert that such theories are not scientific because they are not naturalistic simply assumes the point at issue. Of course intelligent design is not wholly naturalistic, but why does that make it unscientific? What noncircular reason can be given for this assertion? What independent criterion of method demonstrates the inferior scientific status of a nonnaturalistic explanation? We have seen that "must explain via law" does not. What does?

Unobservables and testability. At this point evolutionary demarcationists must offer other demarcation criteria. One that appears frequently both in conversation and in print finds expression as follows: "Miracles are unscientific because they cannot be studied empirically.[62] Design invokes miraculous events; therefore design is unscientific. Moreover, since miraculous events can't be studied empirically, they can't be tested.[63] Since scientific theories must be testable, design is, again, not scientific." Molecular biologist Fred Grinnell has argued, for example, that intelligent design cannot be a scientific concept because if something "can't be measured, or counted, or photographed, it can't be science".[64] Gerald Skoog amplifies this concern: "The claim that life is the result of a design created by an intelligent cause can not be tested and is not within the realm of science."[65] This reasoning was invoked in a 1993 case at San Francisco State University as a justification for removing Professor Dean Kenyon from his classroom. Kenyon is a biophysicist who has embraced intelligent design after years of work on chemical evolution. Some of his critics at SFSU argued that his theory fails to qualify as scientific because it refers to an unseen Designer that cannot be tested or, as Eugenie Scott said, "You can't use supernatural explanations because you can't put an omnipotent deity in a test tube. As soon as creationists invent a 'theo-meter' maybe we could test for miraculous intervention."[66]

The essence of these arguments seems to be that the unobservable character of a designing agent renders it inaccessible to empirical investigation and thus precludes the possibility of testing any theory of design. Thus the criterion of demarcation employed here conjoins "observability and testability". Both are asserted as necessary to scientific status, and the converse of one (unobservability) is asserted to preclude the possibility of the other (testability).

It turns out, however, that both parts of this formula fail. First, observability and testability are not both necessary to scientific status, because observability at least is not necessary to scientific status, as theoretical physics has abundantly demonstrated. Many entities and events cannot be directly observed or studied—in practice or in principle. The postulation of such entities is no less the product of scientific inquiry for that. Many sciences are in fact directly charged with the job of inferring the unobservable from the observable. Forces, fields, atoms, quarks, past events, mental states, subsurface geological features, molecular biological structures—all are unobservables inferred from observable phenomena. Nevertheless, most are unambiguously the result of scientific inquiry.

Second, unobservability does not preclude testability: claims about unobservables are routinely tested in science indirectly against observable phenomena. That is, the existence of unobservable entities is established by testing the explanatory power that would result if a given hypothetical entity (that is, an unobservable) were accepted as actual. This process usually involves some assessment of the established or theoretically plausible causal powers of a given unobservable entity. In any case, many scientific theories must be evaluated indirectly by comparing their explanatory power against competing hypotheses.

During the race to elucidate the structure of the genetic molecule, both a double helix and a triple helix were considered, since both could explain the photographic images produced via X-ray crystallography.[67] While neither structure could be observed (even indirectly through a microscope), the double helix of Watson and Crick eventually won out because it could explain other observations that the triple helix could not. The inference to one unobservable structure—the double helix—was accepted because it was judged to possess a greater explanatory power than its competitors with respect to a variety of relevant observations. Such attempts to infer to the best explanation, where the explanation presupposes the reality of an unobservable entity, occur frequently in many fields already regarded as scientific, including physics, geology, geophysics, molecular biology, genetics, physical chemistry, cosmology, psychology, and, of course, evolutionary biology.

The prevalence of unobservables in such fields raises difficulties for defenders of descent who would use observability criteria to disqualify design. Darwinists have long defended the apparently unfalsifiable nature of their theoretical claims by reminding critics that many of the creative processes to which they refer occur at rates too slow to observe. Further, the core historical commitment of evolutionary theory—that present species are related by common ancestry—has an epistemological character that is very similar to many present design theories. The transitional life forms that ostensibly occupy the nodes on Darwin's branching tree of life are unobservable, just as the postulated past activity of a Designer is unobservable.[68] Transitional life forms are theoretical postulations that make possible evolutionary accounts of present biological data. An unobservable designing agent is, similarly, postulated to explain features of life such as its information content and irreducible complexity. Darwinian transitional, neo-Darwinian mutational events, punctuationalism's "rapid branching" events, the past action of a designing agent—none of these is directly observable. With respect to direct observability, each of these theoretical entities is equivalent.

Each is roughly equivalent with respect to testability as well. Origins theories generally must make assertions about what happened in the past to cause present features of the universe (or the universe itself) to arise. They must reconstruct unobservable causal events from present clues or evidences. Positivistic methods of testing, therefore, that depend upon direct verification or repeated observation of cause-effect relationships have little relevance to origins theories, as Darwin himself understood. Though he complained repeatedly about the creationist failure to meet the *vera causa* criterion —a nineteenth-century methodological principle that favored theories postulating observed causes—he chafed at the application of rigid positivistic standards to his own theory. As he complained to Joseph Hooker: "I am actually weary of telling people that I do not pretend to adduce *direct* evidence of one species changing into another, but that I believe that this view in the main is correct because so many phenomena can be thus grouped and *explained*"[69] (emphasis added).

Indeed, Darwin insisted that direct modes of testing were wholly irrelevant to evaluating theories of origins. Nevertheless, he did believe that critical tests could be achieved via indirect means. As he stated elsewhere: "This hypothesis [common descent] must be tested . . . by trying to see whether it explains several large and independent classes of facts; such as the geological succession of organic beings, their distribution in past and present times, and their mutual affinities and homologies."[70] For Darwin the unobservability of past events and processes did not mean that origins theories are untestable. Instead, such theories may be evaluated and tested indirectly by the assessment of their explanatory power with respect to a variety of relevant data or "classes of facts".

Nevertheless, if this is so it is difficult to see why the unobservability of a Designer would necessarily preclude the testability of such a postulation. Though Darwin would not have agreed, the basis of his methodological defense of descent seems to imply the possibility of a testable theory of design, since the past action of an unobservable agent could have empirical consequences in the present just as an unobservable genealogical connection between organisms does. Indeed, Darwin himself tacitly acknowledged the testability of design by his own attempts to expose the empirical inadequacy of competing creationist theories. Though Darwin rejected many creationist explanations as unscientific in principle, he attempted to show that others were incapable of explaining certain facts of biology.[71] Thus sometimes he treated creationism as a serious scientific competitor lacking explanatory power; at other times he dismissed it as unscientific by definition.

Recent evolutionary demarcationists have contradicted themselves in the same way. The quotation cited earlier from Gerald Skoog ("The claim that life is the result of a design created by an intelligent cause can not be tested and is not within the realm of science") was followed in the same paragraph by the statement "Observations of the natural world also make these dicta [concerning the theory of intelligent design] suspect."[72] Yet clearly something cannot be both untestable in principle and subject to refutation by empirical observations.

The preceding considerations suggest that neither evolutionary descent with modification nor intelligent design is ultimately untestable. Instead, both theories seem testable indirectly, as Darwin explained of descent, by a comparison of their explanatory power with that of their competitors. As Philip Kitcher—no friend of creationism—has acknowledged, the presence of unobservable elements in theories, even ones involving an unobservable Designer, does not mean that such theories cannot be evaluated empirically. He writes, "Even postulating an unobserved Creator need be no more unscientific than postulating unobserved particles. What matters is the character of the proposals and the ways in which they are articulated and defended."[73]

Thus an unexpected equivalence emerges when design and descent are evaluated against their ability to meet specific demarcation criteria. The demand that the theoretical entities necessary to origins theories must be directly observable if they are to be considered testable and scientific would, if applied universally and disinterestedly, require the exclusion not only of design but also of descent. Those who insist on the joint criteria of observability and testability, conceived in a positivistic sense, promulgate a definition of correct science that evolutionary theory manifestly cannot meet. If, however, a less severe standard of testability is allowed, the original reason for excluding design evaporates. Here an analysis of specific attempts to apply demarcation criteria against design actually demonstrates a methodological equivalence between design and descent.

Other demarcation criteria. I claim that a similar equivalence between design and descent will emerge from an analysis of each of the other criteria—(d) through (h)—listed above.[74] Falsification, for example, in addition to the problems mentioned in part 1, seems an especially problematic standard to apply to origins theories. So does prediction. Origins theories must necessarily offer ex post facto reconstructions. They therefore do not make predictions in any strong sense. The somewhat artificial "predictions" that origins theories do make about, for example, what evidence one ought to find if a given theory is true are singularly difficult to falsify since, as evolutionary paleontologists often explain, "the absence of evidence is no evidence of absence."[75]

Similarly, the requirement that a scientific theory must provide a causal mechanism fails to provide a metaphysically neutral standard of demarcation for several reasons. First, as we have already noted, many theories in science are not mechanistic theories. Many theories that explicate what regularly happens in nature either do not or need not explain why those

174 STEPHEN C. MEYER

phenomena occur mechanically. Newton's universal law of gravitation was no less a scientific theory because Newton failed—indeed refused—to postulate a mechanistic cause for the regular pattern of attraction his law described. Also, as noted earlier, many historical theories about *what* happened in the past may stand on their own without any mechanistic theory about *how* the events to which such theories attest could have occurred. The theory of common descent is generally regarded as a scientific theory even though scientists have not agreed on a completely adequate mechanism to explain how transmutation between lines of descent can be achieved. In the same way, there seems little justification for asserting that the theory of continental drift became scientific only after the advent of plate tectonics. While the mechanism provided by plate tectonics certainly helped render continental drift a more persuasive theory,[76] it was nevertheless not strictly necessary to know the mechanism by which continental drift *occurs* (1) to know or theorize that drift *had occurred* or (2) to regard the continental drift theory as scientific.

Yet one might concede that causal mechanisms are not required in all scientific contexts but deny that origins research is such a context. One might argue that since origins theories necessarily attempt to offer causal explanations, and since design admittedly attempts to explain the origin of life or major taxonomic groups, its failure to offer a mechanism disqualifies it as an adequate theory of origins.

But this argument has difficulties as well. First, an advocate of design could concede that his theory does not provide a complete causal explanation of how life originated without forfeiting scientific status for the theory. Present clues and evidences might convince some scientists *that* intelligence played a causal role in the design of life, without those same scientists' knowing exactly *how* mind exerts its influence over matter. All that would follow in such a case is that design is an incomplete theory, not that it is an unscientific one (or even an unwarranted one). And such incompleteness is not unique to design theories. Both biological (as just discussed) and chemical evolutionary theories have often provided less than completely adequate causal scenarios. Indeed, most scientific theories of origin are causally incomplete or inadequate in some way.

In any case, asserting mechanism as necessary to the scientific status of origins theories begs the question. In particular, it assumes without justification that all scientifically acceptable causes are *mechanistic* causes. To insist that all causal explanations in science must be mechanistic is to insist that all causal theories must refer only to material entities (or their energetic equivalents). Yet this requirement is merely another expression of the very naturalism whose methodological necessity has been asserted because of ostensibly compelling demarcation arguments. Insofar as the statement "All scientific theories must be mechanistic" *is* a demarcation argument, this requirement is evidently circular. Science, the demarcationist claims, must be mechanistic because it must be naturalistic; it must be naturalistic because otherwise it would violate demarcation standards—in particular, the standard that all scientific theories must be mechanistic.

This argument clearly assumes the point at issue, which is whether or not there are independent—that is, metaphysically neutral—reasons for preferring exclusively materialistic causal explanations of origins over explanations that invoke putatively immaterial entities such as creative intelligence, mind, mental action, divine action, or intelligent design. While philosophical naturalists may not regard the foregoing as real, they certainly cannot deny that such entities could function as causal antecedents if they were.

Thus we return to the central question: What noncircular reason can be offered for prohibiting the postulation of nonmechanistic (for instance, mental or intelligent) causes in scientific origins theories? Simply asserting that such entities may not be considered, whatever the empirical justification for their postulation, clearly does not constitute a justification for an exclusively naturalistic definition of science. Theoretically

there are at least two possible types of causes: mechanistic and
intelligent. The demarcationist has yet to offer a noncircular
reason for excluding the latter type.[77]

Part 3: The Methodological
Character of Historical Science

Let us now turn to a more fundamental reason for the methodological equivalence of design and descent. As stated earlier, the
equivalence of design and descent follows from an understanding of the distinctive logical and methodological character of
the historical sciences. An examination of scientific disciplines
concerned with past events and causes, such as evolutionary biology, historical geology, and archaeology, reveals a distinctive
pattern of inquiry that contrasts markedly with nonhistorical
sciences such as branches of chemistry, physics, or biology
that are concerned primarily with the discovery and explication of general phenomena. This section will show that both
design and descent do, or could, instantiate this distinctive
historical pattern of scientific investigation. In other words, a
fundamental methodological equivalence between design and
descent derives from a common concern with history—that is,
with historical questions, historical inferences, and historical
explanations.

We can see this historical concern first by looking at why
the demarcation arguments analyzed earlier fail. Consider, for
example, the assertion that to be scientific one must explain by
reference to natural law. To insist that "science must explain
by natural law" betrays much confusion—about the alleged
universality of explanation in science, about the necessary role
of laws in explanations, and about the distinction between laws
and causes. But fundamentally this demarcation criterion fails
to do the work required of it by evolutionary writers because
it ignores the fact that some scientific disciplines ("historical",
according to my lexicon) seek to explain events or data not

primarily by reference to laws but by reference to past causal events or sequences of events—what might be called "causal histories". Since natural laws are not necessary to such activity, the demarcation criterion "must explain by natural law" cannot be used to distinguish between two competing programs of historical scientific research, whether evolutionary or otherwise.

Next consider the idea that scientific theories must not postulate unverifiable or unobservable entities. Certainly this criterion is untenable in light of many fields, not the least of which is modern physics. Yet it is completely irrelevant to historical study almost in principle. All historical theories depend on what C. S. Peirce called "abductive inferences".[78] Such inferences frequently posit unobservable past events in order to explain present phenomena, facts, or clues. Making a claim about history nearly always involves postulating, invoking, or inferring an unobservable event or entity that cannot be studied directly. The attempt to distinguish the methodological merit of competing origins theories on the basis of unobservables therefore seems quite misguided and futile.

Finally, consider the claim that to be scientific a theory must be testable. As we saw above, neither design nor descent can meet standards of testability that require strict verifiability. I have also emphasized that neither can meet standards of testability that depend on notions of repeatability. Yet both can meet alternate standards of testability, such as inference to the best explanation or "consilience", that involve notions of comparative explanatory power. This equivalence was suggested again from the historical nature of the claims that design and evolutionary theorists make. Like other historical theorists, both make claims about events they believe occurred in the past that cannot be directly verified and may never recur. Yet like other historical theories, these theories can be tested after the fact by reference to their comparative explanatory power. To impose stricter standards ignores the limitations inherent in all historical inquiry and thus again fails to provide grounds for distinguishing the status of competing historical or origins theories.

So the evolutionary demarcation arguments above seem to fail in part because they attempt to impose (as normative) criteria of method that ignore the historical character of origins research. Indeed, each one of the demarcationist arguments listed above fails because it overlooks a specific characteristic of the historical sciences. But what are these characteristics? And could *they* provide grounds for distinguishing the scientific, or at least methodological, status of design and descent?

The nature of historical science. Answering these questions will require briefly summarizing the results of my doctoral research on the logical and methodological features of the historical sciences.[79] Through that research I have identified three general features of historical scientific disciplines. These features derive from a concern to reconstruct the past and to explain the present by reference to the past. They distinguish disciplines motivated by historical concerns from disciplines motivated by a concern to discover, classify, or explain unchanging laws and properties of nature. These latter disciplines may be called "inductive" or "nomological" (from the Greek word *nomos*, for law); the former type may be called "historical".[80] I contend that historical sciences generally can be distinguished from nonhistorical scientific disciplines by virtue of the three following features.

1. The historical interest or questions motivating their practitioners: Those in the historical sciences generally seek to answer questions of the form "What happened?" or "What caused this event or that natural feature to arise?" On the other hand, those in the nomological or inductive sciences generally address questions of the form "How does nature normally operate or function?"

2. The distinctively historical types of inference used: The historical sciences use inferences with a distinctive logical form. Unlike many nonhistorical disciplines, which typically attempt to infer generalizations or laws from particular facts, historical

sciences make what C. S. Peirce has called "abductive inferences" in order to infer a past event from a present fact or clue.
These inferences have also been called "retrodictive" because
they are temporally asymmetric—that is, they seek to reconstruct past conditions or causes from present facts or clues. For
example, detectives[81] use abductive or retrodictive inferences
to reconstruct the circumstances of a crime after the fact. In so
doing they function as historical scientists. As Gould has put
it, the historical scientist proceeds by "inferring history from
its results".[82]

3. The distinctively historical types of explanations used: In
the historical sciences one finds causal explanations of particular events, not nomological descriptions or theories of general
phenomena. In historical explanations, past causal events, not
laws, do the primary explanatory work. The explanations cited
earlier of the Himalayan orogeny and the beginning of World
War I exemplify such historical explanations.[83]

In addition, the historical sciences share with many other
types of science a fourth feature.

4. Indirect methods of testing such as inference to the best
explanation: As discussed earlier, many disciplines cannot test
theories by direct observation, prediction, or repeated experiment. Instead, testing must be done indirectly through comparison of the explanatory power of competing theories.

Descent as historical science. Enough has been said previously
—about the function of common descent as an explanatory
causal history, the retrodictive character of Darwin's inference
of common descent, and his use of indirect methods of theory
evaluation—to suggest that evolutionary research programs
conform closely to the general methodological pattern of the
historical sciences. But a few additional observations may make
this connection more explicit.

With respect to the first characteristic of historical science enumerated above (historical motive or purpose), Darwin
clearly was motivated by such a purpose. One of Darwin's primary goals in the *Origin of Species* was to establish a historical

point[84]—namely, that species had not originated independently but had derived via transmutation from one or very few common ancestors. Indeed, Darwin sought to show that the history of life resembled a single, continuous, branching tree, with the first and simplest living forms represented by the base of a tree and the great diversity of more complex forms, both past and present, represented by the connecting branches. This picture of biological history contrasted markedly with that of his creationist opponents, who envisioned the history of life as an array of parallel (nonconvergent) lines of descent. Darwin's (perhaps primary) purpose in the *Origin of Species* was to argue for this continuous view of life's history as opposed to the discontinuous view favored by his creationist opponents.

Thus he would repeatedly explicate his priorities in such a way as to show the primacy of his concern to demonstrate the historical thesis of common descent, even over his concern to establish the efficacy of his proposed mechanism, natural selection. He himself tells us what he had in mind: "I had two distinct objects in view; *firstly* to shew that species had not been separately created [that is, that they had evolved from common ancestors], and *second*, that natural selection had been the chief agent of change"[85] (emphasis added).

Similarly, at the close of his chapter 13 Darwin states the priorities of his argument by concluding: "The several classes of facts which have been considered . . . proclaim so plainly that the innumerable species, genera, and families with which the world is peopled are all *descended* . . . from common parents and have been modified in the course of descent, that I should without hesitation adopt this view, *even if* it were unsupported by other facts or arguments"[86] (emphasis added).

Not only was Darwin motivated by a historical purpose, but he also used (concerning feature 2 above) a characteristically historical mode of reasoning. As Gould has argued so persuasively, Darwin used historical inferences. Beginning in the middle of his chapter on the "Geological Succession of Organic Beings" and continuing through his next three chapters, Darwin offered a series of arguments to support his historical claim of common descent.[87] These arguments are instances of retrodictive or abductive reasoning. In each case, extant evidence from the fossil record, comparative anatomy, embryology, and biogeography were used as clues from which to infer a pattern of past biohistorical events. Notice, for example, the language Darwin uses in his argument from vestigial structures: "Rudimentary organs may be compared with the letters in a word, still retained in the spelling but become useless in the pronunciation, but *which serve as a clue in seeking for its derivation.*"[88]

Notice, too, the temporally asymmetric character of each of the inferences he employs: "The several *classes of facts* which have been considered . . . proclaim so plainly that the innumerable species, genera, and families with which the world is peopled are all *descended*, each within its own class or group, *from common parents.*"[89] As Gould has written, Darwin used a method of "inferring history from its results".[90]

Darwin not only inferred a historical past, but (with respect to feature 3 above) he also formulated historical explanations. Indeed, a reciprocal relationship exists between historical inferences and explanations. Historical scientists will often seek to infer causal antecedents that, if true, would explain the widest class of relevant data. The causal past inferred on the basis of its potential to explain will often serve, when accepted, as an explanation. Darwin repeatedly argued that the supposition that all organisms descended from common parents should be accepted because it "explains several large and independent classes of facts".[91] Moreover, common descent (and the past events implied by it) served as a *causal* explanation for Darwin. He refers to "propinquity of descent" as "*the only known cause* of the similarity of organic beings".[92] Elsewhere he refers to common descent or "propinquity of descent" as the *vera causa* (or true cause) of organic similarity.[93] By inferring descent as a past cause, Darwin constructed a historical explanation in which a pattern of past events did the primary explanatory work in relation to the facts of biogeography, fossil progression, homology and so on. As Gould has put it, the *Origin of Species* makes "the claim that *history* stands as the coordinating reason for relationships among organisms".[94]

The explanatory function of antecedent events and causal histories is perhaps even more readily apparent in the work of many chemical evolutionary theorists. Alexander Oparin, Russian scientist and father of modern origin-of-life research, formulated detailed causal histories involving a sequence of hypothetical past events to explain how life emerged in its present form.[95] The formulation of these "scenarios", as they are called in origin-of-life biology, has remained an important part of origin-of-life studies to the present.[96] Thus evolutionary biologists employ not only historical inferences but also historical explanations in which past causal events, or patterns thereof, serve to explain the origin of present facts.

As already discussed, Darwin also (with respect to feature 4 above) employed a method of indirect testing of his theory by assessing its relative explanatory power. Recall his statement that "this hypothesis [that is, common descent] must be tested . . . by trying to see whether it explains several large and independent classes of facts."[97] He makes this indirect and comparative method of testing even more explicit in a letter to Asa Gray:

> I . . . test this hypothesis [common descent] by comparison with as many general and pretty well-established propositions as I can find—in geographical distribution, geological history, affinities &c., &c. And it seems to me that, *supposing* that such a hypothesis were to explain such general propositions, we ought, in accordance with the common way of following all science, to admit it till some *better* hypothesis be found out [emphasis added].[98]

Design as historical science. The foregoing suggests that evolutionary biology, or at least Darwin's version of it, does conform to the pattern of inquiry described above as historically scientific. To show that design and descent are methodologically equivalent with respect to the historical mode of inquiry outlined above, it now remains to show that a design argument or theory could exemplify this same historical pattern of inquiry.

In the case of feature 1, this equivalence is quite obvious. As just noted, a clear logical distinction exists between questions of the form "How does nature normally operate or function?" and those of the form "How did this or that natural feature arise?" or "What caused this or that event to occur?" Those who postulate the past activity of an intelligent Designer do so as an answer, or partial answer, to questions of the latter historical type. Whatever the evidential merits or liabilities of design theories, such theories undoubtedly represent attempts to answer questions about what caused certain features in the natural world to come into existence. With respect to an interest in origins questions, design and descent are clearly equivalent.

Design and descent are also equivalent with respect to feature 2. Inferences to intelligent design are clearly abductive and retrodictive. They seek to infer a past unobservable cause (an instance of creative mental action or agency) from present facts or clues in the natural world, such as the informational content in DNA, the irreducible complexity of molecular machines, the hierarchical top-down pattern of appearance in the fossil record, and the fine tuning of physical laws and constants.[99] Moreover, just as Darwin sought to strengthen the retrodictive inferences that he made by showing that many facts or classes of facts could be explained on the supposition of common descent, so too may proponents of design seek to muster a wide variety of clues to demonstrate the explanatory power of their theory.

With respect to feature 3, design inferences, once made, may also serve as causal explanations. The same reciprocal relationship between inference and explanation that exists in arguments for descent can exist in arguments for design. Thus, as noted, an inference to intelligent design may gain support because it could, if accepted, explain many diverse classes of facts. Clearly, once adopted it will provide corresponding explanatory resources. Moreover, theories of design involving

184 STEPHEN C. MEYER

the special creative act of an agent conceptualize that act as a causal event,[100] albeit involving mental rather than purely physical antecedents. Indeed, design theories—whether posited by young-earth Genesis literalists, old-earth progressive creationists, theistic macromutationalists, or religiously agnostic biologists—refer to antecedent causal events or express some kind of causal scenario just as, for example, chemical evolutionary theories do. As a matter of method, advocates of design and descent alike seek to postulate antecedent causal events or event scenarios in order to explain the origin of present phenomena. With respect to feature 3, design and descent again appear methodologically equivalent.

Much has already been said to suggest that with respect to feature 4 design may be tested indirectly in the same way as descent. Certainly, advocates of design may seek to test their ideas as Darwin did—against a wide class of relevant facts and by comparing the explanatory power of their hypotheses against those of competitors. Indeed, many biologists who favor design now make their case for it on the basis of its ability to explain the same evidences that descent can as well as some that descent allegedly cannot (such as the presence of specified complexity or information content in DNA).[101]

Thus design and descent again seem methodologically equivalent. Both seek to answer characteristically historical questions; both rely upon abductive inferences; both postulate antecedent causal events or scenarios as explanations of present data; and both are tested indirectly by comparing their explanatory power against that of competing theories.

A theory of everything? Yet before one is willing to concede this methodological equivalence, one might demand to know whether design can really function as a valid explanation without trivializing scientific inquiry. The worry about theories of design concerns, not their explanatory power, but the inability to constrain that power. Would a theory of design leave scientists with nothing to do, since presumably the phrase "God did it" could be invoked as the answer to every scientific question? As David Hull wrote recently, "Scientists have no choice [but to define science as totally naturalistic]. Once they allow reference to God or miraculous forces to explain the first origin of life or the evolution of the human species, they have no way of limiting this sort of explanation."[102] This also finds expression in the familiar theistic worry about "God-of-the-gaps" arguments. So both theists and secularists may worry: "If design is allowed as a (historically) scientific theory, could it not be invoked at every turn as a theoretical panacea, stultifying inquiry as it goes? Might not design become a refuge for the intellectually lazy who have refused to study what nature actually does?"

The distinction between the historical and the nomological helps to show how design can be both appropriate and inappropriate (and thus constrained) depending upon the context of inquiry. That is, this distinction helps to show why the past action of an intelligent agent may serve as a legitimate explanation in the historical sciences, whereas it would not in many nonhistorical scientific contexts.

When scientists address questions of what nature normally does or how one part of nature generally affects another, any reference to the particular action of agents becomes inappropriate because it fails to address the question motivating the inquiry. Consider the question: "How does atmospheric pressure affect crystal growth?" To state "Crystals were designed by a creative intelligence" (or, for that matter, "Crystals evolved via natural processes") fails to answer the question. Here appropriate answers are necessarily both naturalistic and nomological because the question asks how one part of nature generally affects another. Yet a naturalistic answer is necessary only because of the focus of the question. Inductive sciences typically seek to establish general causal or descriptive *relationships* (laws), whereas historical sciences typically infer particular past causal *events*. To propose the action of agency (as an event in space and time) when a descriptive or causal law is required fails to address the challenge of nomological inquiry. To answer "God created it" to a geologist who inquires about the stress/strain relationship of a particular rock type or to a cell biologist inquiring about how a given protein normally binds to the cytoskeleton is contextually inappropriate. Neither divine nor human action qualifies as a law. Such answers do not violate the rules of science as much as they violate common sense considerations of context. They do stultify inquiry, but only because they miss the point of a particular type of inquiry altogether.

It not does follow, however, that references to agency are necessarily inappropriate when reconstructing a causal history —when attempting to answer questions about how a particular feature in the natural world (or the universe itself) arose. First, classical examples of inappropriate postulations of divine activity (that is, God-of-the-gaps arguments) occur almost exclusively in the inductive or nomological sciences, as Newton's ill-fated use of agency to provide a more accurate description of planetary motion suggests.[103] Secondly, the action of agents is routinely invoked to account for the origin of features or events within the natural world. Forensic science, history, and archeology, for example, all sometimes postulate the past activity of human agents to account for the emergence of particular objects or events. Several such fields suggest a clear precedent for inferring the past causal activity of intelligent agents within the historical sciences. (Imagine the absurdity of someone claiming that scientific method had been violated by the archeologist who first inferred that French cave paintings had been produced by human beings rather than by natural forces such as wind and erosion.)

There is another more fundamental reason why postulating the past action of agency can be appropriate in the historical sciences: historical explanations require the postulation of antecedent causal events; they do not seek to infer laws.[104] To offer past agency as part of an origins scenario or explanation is, therefore, contextually appropriate because the type of theoretical entity provided corresponds to the type required by historical explanations. Simply put, past agency is a causal event. Agency, therefore, whether seen or unseen, may serve as a contextually appropriate theoretical entity in a historical explanation, even if it could not do so in a nomological or inductive theory. Mental action may be a causal event, even if it is not a law.

In any case, postulations of design are constrained by theoretical competition. The plausibility of historical theories must be adjudicated against background information about the causal powers and proclivities of both nature and agency.[105] Intelligent design can be offered, therefore, as a necessary or best historical explanation only when available naturalistic processes seem incapable of producing the *explanandum* effect, and when intelligence is known to be capable of, and thought inclined to, produce it. Thus, modern scientific advocates of intelligent design such as Charles Thaxton, Walter Bradley, Dean Kenyon, Michael Behe, and William Dembski insist that they postulate antecedent intelligent activity, not because of what we do not know, but because of what we do know about what is, and is not, capable of producing, for example, "information content" (Meyer, Thaxton and Bradley, and Kenyon),[106] "small probability specifications" (Dembski),[107] or "irreducible complexity" (Behe).[108] Conversely, there are many effects that do not, based upon our present background knowledge of causal powers, suggest design as either a necessary or best explanation.

An example may help to illustrate this. In the wooded neighborhood where I live near Whitworth College there are many pine trees. Every Friday morning I notice a group of pine cones piled neatly on a blue tarp next to my neighbor's curbside garbage cans. I know that Friday is trash day and that my elderly neighbor detests pine cones, pine needles, and other debris on his lawn. Given this background information, I infer that he has intentionally piled the pine cones on his tarp by design. While it is true that I have occasionally seen a few pine cones clustered together under his trees, I have never seen so many piled so neatly "in the wild". Nor do I think it likely that natural causes could have removed my neighbor's tarp from his garage and positioned it next to his garbage cans (and under the cones) without assistance. Wind, rain, and gravity may be powerful, but they are not that smart. Thus, on the basis of my background knowledge about the capabilities of nature and agents (and in this case, the proclivities of my neighbor), I infer that personal agency—intelligent design—has played a causal role in the assembly of the pile next door.[109] Indeed, I make a similar inference every fall when I walk on campus to see that some mysterious agency has spelled out the names of the freshmen dormitories—"Stewart Hall", "MacMillian Hall", and so on—with mounds of still more pine cones on the lawns of these residence halls.

Nevertheless, I do not always infer intelligent design as the cause of every phenomenon. My own lawn is usually covered with pine cones in a haphazard fashion. While it is possible that the pine cones in my yard assumed their configuration as the result of a personal agent, this seems quite unlikely to me. First, I doubt that anyone would waste his time placing pine cones in my yard in such a random arrangement. Second, I have witnessed pine cones falling and producing such random scatters many times. Thus, the distribution of pine cones on my lawn seems best explained by a combination of natural factors: wind, rain, gravity, the position of the trees, the slope of the yard, the length of the grass, and so on. Similarly, the configuration of cones in another neighbor's yard, though arranged in a less random way in two distinct clusters, also seems to reflect purely natural causes, since each cluster lies just beneath one of the two solitary pines in the adjacent yard.

In both these cases—that is, where the cones are fairly randomly scattered and where they are clustered in a more orderly way—intelligent design does not seem the best explanation, even though agents are capable of producing such effects. Instead, in the absence of good reasons to suspect that agents *would* want to produce these effects and some sign that one did, it seems more probable to attribute these effects to natural processes with proven causal efficacy. Moreover, as Dembski has shown, neither low probability events nor high probability events allow intelligent design to be unambiguously detected. Instead, intelligent design can be unambiguously detected only in *specified* events of very small probability.[110] The pine cones on the tarp and especially those spelling messages on the dormitory lawns provide good examples of small probability events that are specified. In the latter case, the improbable arrangement of the cones is specified by an alphabet convention in order to achieve communication. Since contemporary design theorists do not assert only that intelligent design has occurred but that the effects of design are unambiguously detectable in certain natural features such as specified complexity or information encoded along the spine of the DNA molecule, their theoretical claims are anything but vacuous or trivial.

Indeed, if design theorists are correct, design cannot be inferred for every effect, even if intelligent design is a possible cause of all effects. Because intelligent agents, and presumably the Divine Agent, have causal powers that nature does not have, intelligent design may always be a possible explanation. Nevertheless, possible explanations are not necessarily the best explanations. Intelligent design is not always the best explanation for a variety of reasons. Human action or special (that is, detectable) divine action may not have played a causal role in certain natural events; intelligent design, whether human or divine, may not always be detectable even when it has played a causal role; natural objects and processes have real causal powers (even for theists who accept God's sustaining governance of nature) that may be clearly evident in a given phenomenon. Thus, at least for those scientists who seek the best explanations, intelligent design cannot be invoked as a theory of everything. It may function as a possible theory of everything, but it can function as the best explanation or best theory of only some things. Intelligent design need be neither vacuous nor unconstrained.

Further, postulations of intelligent design are constrained by background assumptions about the proclivities of potential designing agents, both human and divine. In particular they are constrained by assumptions about the assumed character and inclinations of God. Most biblical theists, for example, assume that God acts in at least two ways: (1) through the natural regularities or laws that he upholds and sustains through his invisible power and (2) through more dramatic, discernible, and discrete actions at particular points in time. Because theists assume that the second mode of divine action is by far the more rare, and usually associated with the accomplishment of some particular divine purpose on behalf of human beings (for example, creation or redemption), theists assume that divine action of the second variety will be unlikely as an explanation of most particular events. It might be the case that the windstorm that blew the scales of justice off of the Old Bailey in London in 1987 was a special act of God, but most theists would—in the absence of any discernible redemptive import associated with the event—tend to regard it as part of the ordinary (albeit God-governed) concourse of nature. Theists generally approach their study of nature with a set of background assumptions that would lead them to regard most hypotheses of special divine action as unlikely, though not completely impossible. As such, theism itself constrains design inferences. Theistic background assumptions would generally allow consideration of special divine action as the best or most likely explanation for a particular event only when it seemed empirically warranted and theologically plausible. Nevertheless, given a biblical (though not necessarily literalist) understanding of creation and sufficient empirical justification, there is no reason to believe that both these conditions could not be met in some cases, as with, for example, explanations of the origin of life, human consciousness, or the universe.

An example of theological plausibility functioning to limit design hypotheses can be found by examining the reception of Newton's famous postulation of special divine intervention to stabilize the orbital motion in the solar system. Newton postulated the periodic and special intervention of God to correct for an apparently accumulating instability in the orbits of the outer planets (Jupiter and Saturn) within the solar system. While this episode is often cited to illustrate why divine action or design can never be considered as a scientific explanation, it actually illustrates a more subtle point: how such inferences were constrained by considerations of theological plausibility.

To many eighteenth-century scientists, Newton's interventionist theory seemed ill-formed and unlikely not because it contradicted an inviolate methodological convention, as has often been asserted.[111] Newton himself made highly regarded design arguments in other contexts and believed gravitation was caused by constant spirit action.[112] Instead, Newton's argument for angelic action was rejected because it seemed both theologically unlikely (given prevailing background assumptions about how God interacts with nature and given the nomological context of inquiry) and less elegant than the explanation that Laplace would later offer in the 1770s.

The theistic research program of Newton's day assumed that the regularity and universality of natural laws reflected the ordered mind and sovereign power of the Creator. Kepler and Newton both wanted to use science to demonstrate this. To hypothesize as Newton did that divine gerrymandering was required to maintain the orbital stability of the solar system seemed improbable and ad hoc to *theistic* scientists. It did so because it clearly violated, not a methodological prohibition against reference to divine action, but a fundamental theological background assumption of many scientists at the time —namely, that special or discrete divine action was unlikely and unnecessary where God's *potentia ordinata*, his regular orderly power, was sufficient and already at work.[113] Thus, when Laplace later demonstrated the stability of the planetary system by showing that orbital perturbations oscillated within fixed quantifiable limits,[114] he "saved" the very regularity of celestial mechanics that was the triumph of the theistic research program initiated by Kepler and later advanced by Newton himself via the theory of universal gravitation.

The preceding considerations suggest that allowing the design hypothesis as the best explanation for some events in the history of the cosmos will not cause science to come grinding to a halt. While design does have the required features of some scientific (historical) explanations, it cannot be invoked appropriately in all scientific contexts. Furthermore, because effective postulations of design are constrained by empirical considerations of causal precedence and adequacy, and by extraevidential considerations such as simplicity and theological plausibility, concerns about design theory functioning as a "theory of everything" or "providing cover for ignorance" or "putting scientists out of work" can be shown to be unfounded.[115] Many important scientific questions would remain to be answered if one adopted a theory of design. Indeed, all questions about how nature normally operates without the special assistance of divine agency remain unaffected by whatever view of origins one adopts. And that, perhaps, is yet another equivalence between design and descent.[116]

Conclusion: Toward a
Scientific Theory of Creation

So what should we make of these methodological equivalencies? Can there be a scientific theory of intelligent design?

At the very least it seems we can conclude that we have not yet encountered any good reason in principle to *exclude* design from science. Design seems to be just as scientific (or unscientific) as its naturalistic competitors when judged according to the methodological criteria examined above. Moreover, if the antidemarcationists are correct, our lack of universal demarcation criteria implies there cannot be a negative a priori case against the scientific status of design—precisely because there is not an agreed standard as to what constitutes the properly scientific. To say that some discipline or activity qualifies as scientific is to imply the existence of a standard by which the scientific status of an activity or discipline can be assessed or adjudicated. If no such standard presently exists, then nothing positive (or negative) can be said about the scientific status of intelligent design (or any other theory, for that matter).

But there is another approach that can be taken to the question. If (1) there exists a distinctively historical pattern of inquiry, and (2) a program of origins research committed to design theory could or does instantiate that pattern, and (3) many other fields such as evolutionary biology also instantiate that pattern, and (4) these other fields are already regarded by convention as science, there can be a very legitimate if convention-dependent sense in which design may be considered scientific. In other words, the conjunction of the methodological equivalence of design and descent and the existence of a convention that regards theories of descent as scientific implies that design should—by that same convention—be regarded as scientific too. Thus, one might quite legitimately say that both design and descent are historically scientific research programs, since they instantiate the same pattern of inquiry.

Perhaps, however, one just really does not want to call intelligent design a scientific theory. Perhaps one prefers the designation "quasi-scientific historical speculation with strong metaphysical overtones". Fine. Call it what you will, provided the same appellation is applied to other forms of inquiry that have the same methodological and logical character and limitations. In particular, make sure both design and descent are called "quasi-scientific historical speculation with strong metaphysical overtones".

This may all seem very pointless, but that in a way is just the point. As Laudan has argued, the question whether a theory is scientific is really a red herring. What we want to know is not whether a theory is scientific but whether a theory is true or false, well confirmed or not, worthy of our belief or not. One cannot decide the truth of a theory or the warrant for believing a theory to be true by applying a set of abstract criteria that purport to tell in advance how all good scientific theories are constructed or what they will in general look like.

Against method? Now none of the above should be construed to imply that methodology does not matter. The purpose of this essay is not to argue, as Paul Feyerabend does, against method.[117] Methodological standards in science can be important for guiding future inquiry along paths that have been successful in the past. The uniformitarian and/or actualistic method in the historical sciences, for example, has proved a very helpful guide to reconstructing the past, even if it cannot be used as demarcation between science and pseudoscience, and even if some theories constructed according to its guidelines turn out to be false.

Standards of method may also express some minimal logical and epistemic conditions of success—for example, the conditions related to causal explanation.[118] Successful causal explanations must as a condition of logical sufficiency cite more than just a necessary condition of a given outcome.[119] To explain why a given explosion occurred, it will not suffice to note that oxygen was present in the atmosphere; nor can the death of a patient be explained simply by citing the patient's birth, though clearly birth is necessary to death. These cases illustrate how methodological guidelines (whether tacit or explicit) can help eliminate certain (in this case logically) inadequate hypotheses, even if such guidelines cannot be used to define science exhaustively. Methodological anarchism need not result from a rejection of methodological demarcation arguments.

Nevertheless, following methodological criteria and recipes (of any of the preceding types) does not guarantee theoretical success; nor, again, can such recipes be used to define science exhaustively, if for no other reason than the variety of scientific methods that exist. Moreover, methodological recipes can sometimes become fatal to the success of inquiry if they so dictate the content of acceptable theorizing that they automatically eliminate empirically and logically possible explanations or theories.

And this, I believe, has occurred within origins research. The deployment of flawed or metaphysically tendentious demarcation arguments against legitimate theoretical contenders has produced an unjustified confidence in the epistemic standing of much Darwinian dogma, including "the fact of evolution" defined as common descent. If competing hypotheses are eliminated before they are evaluated, remaining theories may acquire an undeserved dominance.

So the question is not whether there can be a scientific theory of design or creation. The question is whether design should be considered as a competing hypothesis alongside descent in serious origins research (call it what you will). Once issues of demarcation are firmly behind us, understood as the red herrings they are, the answer to this question *must* clearly be yes—that is, if origins biology is to have standing as a fully rational enterprise, rather than just a game played according to rules convenient to philosophical materialists.

Naturalism: the only game in town? G. K. Chesterton once said that "behind every double standard lies a single hidden agenda."[120] Advocates of descent have used demarcation arguments to erect double standards against design, suggesting that the real methodological criterion they have in mind is naturalism. Of course for many the equation of science with the strictly materialistic or naturalistic is not at all a hidden agenda. Scientists generally treat "naturalistic" as perhaps the most important feature of their enterprise.[121] Clearly, if naturalism is regarded as a necessary feature of all scientific hypotheses, then design will not be considered a scientific hypothesis.

But must all scientific hypotheses be entirely naturalistic? Must scientific origins theories, in particular, limit themselves to materialistic causes? Thus far none of the arguments advanced in support of a naturalistic definition of science has provided a noncircular justification for such a limitation. Nevertheless, perhaps such arguments are irrelevant. Perhaps scientists should just accept the definition of science that has come down to them. After all, the search for natural causes has served science well. What harm can come from continuing with the status quo? What compelling reasons can be offered for overturning the prohibition against nonnaturalistic explanation in science?

In fact, there are several. First, with respect to origins, defining science as a strictly naturalistic enterprise is metaphysically gratuitous. Consider: It is at least logically possible that a personal agent existed before the appearance of the first life on earth. Further, as Bill Dembski has argued,[122] we do live in the sort of world where knowledge of such an agent could possibly be known or inferred from empirical data. This suggests that it is logically and empirically possible that such an agent (whether divine or otherwise) designed or influenced the origin of life on earth. To insist that postulations of past agency are inherently unscientific in the historical sciences (where the express purpose of such inquiry is to determine what happened in the past) suggests we know that no personal agent could have existed prior to humans. Not only is such an assumption intrinsically unverifiable, it seems entirely gratuitous in the absence of some noncircular account of why science should presuppose metaphysical naturalism.

Second, to exclude by assumption a logically and empirically possible answer to the question motivating historical science seems intellectually and theoretically limiting, especially since no equivalent prohibition exists on the possible nomological relationships that scientists may postulate in nonhistorical sciences. The (historical) question that must be asked about biological origins is not "Which materialistic scenario will prove most adequate?" but "How did life as we know it actually arise on earth?" Since one of the logically and syntactically appropriate answers to this latter question is "Life was designed by an intelligent agent that existed before the advent of humans", it seems rationally stultifying to exclude the design hypothesis without a consideration of all the evidence, including the most current evidence, that might support it.

The a priori exclusion of design diminishes the rationality of origins research in another way. Recent nonpositivistic accounts of scientific rationality suggest that scientific theory evaluation is an inherently comparative enterprise. Notions such as consilience[123] and Peter Lipton's inference to the best explanation[124] discussed above imply the need to compare the explanatory power of competing hypotheses or theories. If this process is subverted by philosophical gerrymandering, the rationality of scientific practice is vitiated. Theories that gain acceptance in artificially constrained competitions can claim to be neither "most probably true" nor "most empirically adequate". Instead such theories can only be considered "most probable or adequate among an artificially limited set of options".

Moreover, where origins are concerned only a limited number of basic research programs are logically possible.[125] (Either brute matter has the capability to arrange itself into higher levels of complexity, or it does not. If it does not, then either some external agency has assisted the arrangement of matter, or matter has always possessed its present arrangement.) The exclusion of one of the logically possible programs of origins research by assumption, therefore, seriously diminishes the significance of any claim to theoretical superiority by advocates of a remaining program. As Phillip Johnson has argued,[126] the use of "methodological rules" to protect Darwinism from theoretical challenge has produced a situation in which Darwinist claims must be regarded as little more than tautologies expressing the deductive consequences of methodological naturalism.

An openness to empirical arguments for design is therefore a necessary condition of a fully rational historical biology. A rational historical biology must not only address the question "Which materialistic or naturalistic evolutionary scenario provides the most adequate explanation of biological complexity?" but also the question "Does a strictly materialistic evolutionary scenario or one involving intelligent agency or some other theory best explain the origin of biological complexity, given all relevant evidence?" To insist otherwise is to insist that materialism holds a metaphysically privileged position. Since there seems no reason to concede that assumption, I see no reason to concede that origins theories must be strictly naturalistic.

NOTES

[1] Creationists such as Louis Agassiz, for example, accepted the notion of successive acts of creation (separated in time) to explain the succession of increasingly complex organisms attested to by the fossils as one moved up the stratigraphic column.

Homology refers to the observed similarity in the structural characteristics of diverse organisms. The bat, the porpoise, the mole, and humans, for example, all possess a pentadactyl (five-pronged) limb. Darwin believed such similarities reflected the fact that each of these diverse organisms shared a common ancestry, while creationist and idealist biologists such as Louis Agassiz and Richard Owen regarded these similarities as having resulted from the use of a similar plan of design by a Creator.

Biogeographical distribution refers to the pattern of distribution of organisms in a geographic region. Darwin believed that the way organisms were distributed geographically demonstrated that they share a common ancestor. Darwin noted, among other things, that the extent to which the Galapagos Island finches differed from each other in several physical characteristics, such as their coloring, their beak size and shape, was related to the distances between different species. His argument persuaded most biologists that the finches did indeed share a common ancestry. While his observations posed a challenge to those nineteenth-century biologists who were committed to the absolute immutability of species, they did not necessarily trouble those creationist biologists who were willing to concede some variation within limits and who postulated separate creation events in different geographic locales. See W. M. Ho, *Methodological Issues in Evolutionary Theory* (Ph.D. diss., University of Oxford, Oxford, England, 1965), pp. 8–68.

[2] Charles Darwin, *The Origin of Species by Means of Natural Selection* (1859, reprint, Harmondsworth: Penguin Books, 1984), p. 334; N. C. Gillespie, *Charles Darwin and the Problem with Creation* (Chicago: University of Chicago Press, 1979), pp. 67–81.

[3] Gillespie, *Darwin*, pp. 1–18, 41–66, 146–56.

[4] The term "positivistic" here refers, not to the "logical positivism" of A. J. Ayer and the Vienna circle, which did not emerge until the 1920s, but to a generic positivism that had begun to influence scientists throughout most of the nineteenth century. As a philosophy of science, nineteenth-century positivism is associated with Auguste Comte. As Gillespie (*Darwin*, pp. 41–66, esp. 54, 167) and many of Darwin's letters and notebooks show (for example, Darwin's letters to Asa Gray and Charles Lyell dated July 20, 1856, and August 2, 1861, respectively, F. Darwin and A. C. Seward, eds., *More Letters of Charles*

Darwin [London: John Murray, 1903], 1:190), Darwin's conception of science was influenced by Comte, who asserted that true science must move beyond references to God (a theological stage) and other unobservable entities (a metaphysical stage) and focus on observable phenomena reducible to laws (positive science). Thus, it is not anachronistic to refer to Darwin as positivistic.

⁵ As Darwin put it, "On the ordinary view of each species having been independently created, we gain no scientific explanation for any one of these facts. We can only say that it has pleased the Creator to command that past and present inhabitants of the world should appear in a certain order [fossil progression] and in certain areas [biogeographical distribution]; that He has impressed upon them the most extraordinary resemblances [homology], and has classed them in groups subordinate to groups. But by such statements we gain no new knowledge; *we do not connect together facts and laws; we explain nothing*." Quoted in Gillespie, *Darwin*, p. 76, emphasis mine.

⁶ Darwin, *Origin*, pp. 201, 430, 453; V. Kavalovski, *The Vera Causa Principle: A Historico-Philosophical Study of a Meta-Theoretical Concept from Newton through Darwin* (Ph.D. diss., University of Chicago, Chicago, Illinois, 1974), pp. 104–29.

⁷ M. Ruse, *Darwinism Defended: A Guide to the Evolution Controversies* (London: Addison-Wesley, 1982), pp. 59, 131–40, 322–24; M. Ruse, "Creation Science Is Not Science", *Science, Technology and Human Values* 7, no. 40 (1982): 72–78; M. Ruse, "A Philosopher's Day in Court", in *But Is It Science? The Philosophical Question in the Creation/Evolution Controversy*, ed. M. Ruse (Buffalo, N.Y.: Prometheus Books, 1988), pp. 13–38; M. Ruse, "Witness Testimony Sheet, *McLean v. Arkansas*", in *Science?* pp. 287–306, esp. 301; M. Ruse, "They're Here!" *Bookwatch Reviews* 2, no. 1 (1989): 4; M. Ruse, "Darwinism: Philosophical Preference, Scientific Inference and Good Research Strategy", in *Darwinism: Science or Philosophy*, ed. J. Buell and V. Hearn (Richardson, Tex.: Foundation for Thought and Ethics, 1994); S. J. Gould, "Genesis and Geology", in *Science and Creationism*, ed. A. Montagu (New York: Oxford University Press, 1984, pp. 126–35); G. S. Stent, "Scientific Creationism: Nemesis of Sociobiology", in Montagu, *Science*, pp. 136–41; R. Root-Bernstein, "On Defining a Scientific Theory: Creationism Considered", in Montagu, *Science*, pp. 64–94; P. L. Quinn, "The Philosopher of Science as Expert Witness", in Ruse, *Science?*, pp. 367–85; L. Laudan, "Science at the Bar—Causes for Concern", in Ruse, *Science?*, pp. 351–55. A. D. Kline, "Theories, Facts and Gods: Philosophical Aspects of the Creation-Evolution Controversy", in *Did the Devil Make Darwin Do It?* ed. D. B. Wilson (Ames: Iowa State University Press, 1983), pp. 37–44; D. J. Futuyma, *Science on Trial: The Case for Evolution* (New York: Pantheon Books, 1983), pp. 161–74; G. Skoog, "A View from the Past", *Bookwatch Reviews* 2 (1989): 1–2; S. J. Gould, "Evolution as Fact and Theory", in Montagu, *Science*, pp. 118–21; P. Kitcher, *Abusing Science: The Case against Creationism* (Cambridge: MIT Press, 1982), pp. 45–54, 126–27, 175–76.

[8] M. Scriven, "Explanation and Prediction in Evolutionary Theory", *Science* 130 (1959): pp. 477–82; P. T. Saunders and M. W. Ho, "Is Neo-Darwinism Falsifiable?—And Does It Matter?" *Nature and System* 4 (1982): 179–96; K. Popper, *Unending Quest* (London: William Collins and Sons, 1974), pp. 167–75.

[9] L. Laudan, "The Demise of the Demarcation Problem," in Ruse, *Science?*, pp. 337–50.

[10] Ibid.

[11] Ruse, *Darwinism*, pp. 59, 131–40, 322–24; Ruse, "Creation Science", pp. 72–78; Ruse, "Philosopher's Day", pp. 13–38; Ruse, "Witness", pp. 287–306, esp. 301; Ruse, "They're Here!"; Ruse, "Darwinism", pp. 21–28; Stent, "Scientific Creationism", pp. 136–41; Root-Bernstein, "Creationism Considered", pp. 64–94; Quinn, "Philosopher", pp. 367–85; Laudan, "Science"; Kline, "Theories", pp. 37–44; Futuyma, *Science*, pp. 161–74; Skoog, "View", pp. 1–2; Gould, "Evolution", pp. 118–21; Kitcher, *Abusing Science*, pp. 45–54, 126–27, 175–76.

[12] Ruse, "Creation Science", pp. 322–24; Stent, "Scientific Creationism", p. 137; Gould, "Evolution", p. 118.

[13] In making this distinction I do not mean to exclude various theories of theistic evolution from consideration as scientific—just the reverse. Such theories vary in content and may be more difficult to classify as theories of either design or descent. Nevertheless, the following classification may prove helpful. Theories that invoke the causal powers of the Divine agent as part of their explanatory framework (that is, where God in some way directs the evolutionary process) can reasonably be considered to be theories of intelligent design, whereas theistic evolutionary theories that do not involve God in their explanatory framework (that is, where God does not in any way direct the evolutionary process but at most upholds the natural law in an undetectable way) can be considered functionally naturalistic and, thus, theories of descent.

[14] James Ebert et al., *Science and Creationism: A View from the National Academy of Science* (Washington, D.C.: National Academy Press, 1987), p. 8.

[15] Laudan, "Demise", pp. 337–50.

[16] Ibid.

[17] O. Gingerich, "The Galileo Affair", *Scientific American*, August 1982, pp. 133–43.

[18] Laudan, "Demise".

[19] Ibid.

[20] Ibid.

[21] I. Lakatos, "Falsification and the Methodology of Scientific Research Programmes", in *Criticism and the Growth of Knowledge*, ed. I. Lakatos and A. Musgrave (Cambridge: Cambridge University Press, 1970), pp. 189–95.

[22] Laudan, "Demise"; Laudan, "Science", p. 354.

[23] This excessive reliance on a philosophical definition of science to circum-

vent the hard work of evaluating specific empirical claims ironically credits the philosophy of science with more power than it possesses. That such appeals to philosophical considerations are typically made by positivist-minded scientists who regard appeals to "philosophy" as anathema only compounds the irony of the demarcationist enterprise. If any demarcating is to be done, it ought to be done by the philosophers of science who specialize in such second-order questions about the definition of science. Yet for reasons specified already, philosophers of science have increasingly spurned this enterprise.

[24] Most who make these demarcation arguments are practicing scientists. Nevertheless, they can be found frequently in the work of the philosopher of science Michael Ruse: *Darwinism*, pp. 59, 131–40, 322–24; "Creation Science", pp. 72–78; "Philosopher's Day", pp. 13–38; "Witness", pp. 287–306, esp. 301; "They're Here!" p. 4; "Darwinism", pp. 1–6.

[25] M. Eger, quoted by J. Buell in "Broaden Science Curriculum", *Dallas Morning News*, March 10, 1989.

[26] Laudan, "Demise", p. 349.

[27] Ruse, "Witness", pp. 287–306; W. R. Overton, "United States District Court Opinion: *McLean v. Arkansas*", in Ruse, *Science?*, pp. 307–31.

[28] It needs to be acknowledged that creationists such as Duane Gish have also employed demarcation arguments against descent: D. Gish, "Creation, Evolution and the Historical Evidence", in Ruse, *Science?*, p. 267.

[29] Ruse, "Witness", p. 301; Ruse, "Philosopher's Day", p. 26; Ruse, "Darwinism", pp. 1–6.

[30] Skoog, "View"; Root-Bernstein, "Creationism Considered", p. 74.

[31] Gould, "Genesis", pp. 129–30; Ruse, "Witness", p. 305; Ebert et al., *Science*, pp. 8–10.

[32] Root-Bernstein, "Creationism Considered," p. 73; Ruse, "Philosopher's Day," p. 28; Ebert et al., *Science*, pp. 8–10.

[33] Kline, "Theories," p. 42; Gould, "Evolution," p. 120; Root-Bernstein, "Creationism Considered," p. 72.

[34] Ruse, *Darwinism*, p. 59; Ruse, "Witness," p. 305; Gould, "Evolution," p. 121; Root-Bernstein, "Creationism Considered", p. 74.

[35] A. Kehoe, "Modern Anti-evolutionism: The Scientific Creationists", in *What Darwin Began*, ed. L. R. Godfrey (Boston: Allyn and Bacon, 1985), pp. 173–80; Ruse, "Witness", p. 305; Ruse, "Philosopher's Day", p. 28; Ebert et al., *Science*, pp. 8–10.

[36] Kitcher, *Abusing Science*, pp. 126–27, 176–77.

[37] Ruse, "Philosopher's Day", pp. 21, 26.

[38] Ibid. One further word of clarification: I am referring to all of the demarcation criteria used in arguments (a)–(h) as methodological criteria. Some of these criteria specify semantic conditions, as noted in my discussion of Laudan's work above. Nevertheless, even these have implications for how scientific theorizing is to be done. To say, for example, that scientific theories must be

falsifiable is also to say that in the process of testing one must, as a matter of method, make a prediction or otherwise state a theory in such a way as to allow its falsification. When I say, therefore, that design and descent are methodologically equivalent, I mean that both approaches to origins are equally capable or incapable of fulfilling the demands of various demarcation criteria, whether strictly methodological, epistemic, or semantic.

[39] Ruse, "Philosopher's Day", pp. 21–26.

[40] Ibid., p. 26; Ruse, "Witness", p. 301.

[41] Ruse, "Darwinism", pp. 1–6; Quinn, "Philosopher", pp. 367–85; Laudan, "Science", pp. 351–55.

[42] By asserting that science must explain by natural law, Ruse is presupposing something called the "covering law" or the "deductive-nomological" view of scientific explanation. The covering-law model was a very popular conception of science during the 1950s and 1960s. It was promulgated primarily by the neopositivist philosopher Carl Hempel. Unfortunately, unsolved problems with the covering-law model of science are legion. C. Hempel, "The Function of General Laws in History", *Journal of Philosophy* 39 (1942): 35–48; G. Graham, *Historical Explanation Reconsidered* (Aberdeen: Aberdeen University Press, 1983), pp. 17–28; Meyer, *Of Clues and Causes: A Methodological Interpretation of Origin of Life Studies* (Ph.D. diss., Cambridge University, 1990), pp. 40–76; W. P. Alston, "The Place of the Explanation of Particular Facts in Science", *Philosophy of Science* 38 (1971): 13–34; M. Scriven, "Explanation", pp. 477–82; M. Scriven, "Truisms as the Grounds for Historical Explanations", in *Theories of History*, ed. P. Gardiner (Glencoe, Ill.: Free Press, 1959), pp. 443–75; M. Scriven, "Causes, Connections and Conditions in History", in *Philosophical Analysis and History*, ed. W. Dray (New York: Harper and Row, 1966), pp. 238–64; M. Mandelbaum, "Historical Explanation: The Problem of Covering Laws", *History Theory* 1 (1961): 229–42; P. Lipton, *Inference to the Best Explanation* (London: Routledge, 1991), pp. 43–46.

[43] The Latin text reads "Hypothesis non fingo." I. Newton, *Isaac Newton's Papers and Letters on Natural Philosophy*, ed. I. Bernard Cohen (Cambridge: Harvard University Press, 1958), p. 302.

[44] Laudan, "Science", p. 334.

[45] Scriven, "Truisms", p. 450; Meyer, *Of Clues*, pp. 40–76.

[46] Alston, "Place"; Meyer, *Of Clues*, pp. 40–75.

[47] Meyer, *Of Clues*, p. 48.

[48] Ibid., pp. 51–56; M. Scriven, "Causation as Explanation", *Nous* 9 (1975): 14; Lipton, *Inference*, pp. 47–81.

[49] Scriven, "Truisms", pp. 446–63, 450. One could, for example, legitimately assert that a particular earthquake caused a bridge to collapse even if all other bridges in the area did not fall and even if all earthquakes do not destroy bridges.

[50] Alston, "Place", pp. 17–24.

[51] Alston makes the same point about laws that state sufficient conditions of a particular outcome as well. Alston (ibid., p. 24) considers the law "Passage of a spark through a mixture of hydrogen and oxygen is sufficient for the formation of water." This, he says, exemplifies a sufficient condition law (hereafter SC). Alston argues that knowing such a law does not alone furnish the scientist with enough information to explain a particular case of water formation, because other sufficient conditions of water formation may have been responsible for the case in question. After all, water forms in a fuel cell without a spark, activating the hydrogen-oxygen combination. Knowing an SC law does not allow one to infer from an instance of the consequent (in this case water formation) that the sufficient condition was antecedently present (in this case a spark in the appropriate gas mixture) unless one *also* knows that the antecedent is the only known sufficient condition of the consequent—that is, unless one knows that the antecedent is both a sufficient and a necessary condition of the consequent. Explaining a case of water formation will require independent evidence that a spark was in fact passed through an appropriate gas mixture (as opposed to some other causal antecedent) prior to the event. As Alston states, we can "not tell from the law itself which of the sufficient conditions is responsible in a particular case". Thus, laws of the SC type do not, without supplementary information, constitute explanations of particular facts. To regard laws and explanations as logically identical is, therefore, again mistaken.

[52] Ibid., p. 17.

[53] Ruse, *Darwinism*, p. 58; Gould, "Evolution", pp. 119–21; M. Ridley, *The Problems of Evolution* (Oxford: Oxford University Press, 1985), p. 15. For a cogent discussion of the different meanings of evolution and the logical independence of the theory of common descent and the various mechanistic theories about how transmutation might occur, see also K. S. Thomson, "The Meanings of Evolution", *American Scientist* 70 (1982): 529–31. Strictly speaking, common descent is an abductive or historical inference, as Ruse himself acknowledges when he speaks of "inferring historical phylogenies" ("Darwinism", p. 7). As defined by C. S. Peirce, abductive inferences attempt to establish past causes by examining their results or effects. As such, it is more accurate to refer to common descent as a theory about facts—that is, a theory about what in fact happened in the past. Unfortunately, such historical theories, and the inferences used to construct them, can be notoriously inconclusive or "underdetermined". As Gould has stated, "Results rarely specify their causes unambiguously" ("The Senseless Signs of History", in *The Panda's Thumb* [New York: Norton, 1984], p. 34). Ho, *Issues*, pp. 8–60; E. Sober, *Reconstructing the Past* (Cambridge: MIT Press, 1988), pp. 1–4.

[54] By "evolution" here they mean continuous morphological change over time in such a way that all, or most all, organisms are related by common ancestry.

[55] Ruse and Gould regard the theory of common descent as so well established as to make it virtually indistinguishable from a "fact". Ruse, *Darwinism*, p. 58; Gould, "Evolution", pp. 119–21.

[56] From the Greek word *nomos*, for law.

[57] Indeed, it is even debatable whether the selection-mutation mechanism of neo-Darwinism can be expressed as a system of laws (i.e., nomologically), though some so-called axiomatists such as Williams and Lloyd have tried. My point here is that whether one regards selection-mutation as a nomological theory or as a mechanistic theory, common descent does not depend on it for its scientific status. The logical and epistemic independence of descent from selection-mutation demonstrates the ability of some theories to explain in the absence of either laws or mechanisms.

[58] Darwin, *Origin*, p. 195.

[59] The untenable nature of Ruse's position is manifest in his own admission that modern evolutionary theory does not meet the demarcation standards that he promulgates elsewhere as normative for his opponents. See, for example, his discussion of population genetics in *Darwinism Defended*, where he acknowledges that "it is probably a mistake to think of modern evolutionists as seeking universal laws at work in every situation" (p. 86).

[60] Ruse, "Darwinism", pp. 1–6; Ruse, "Witness", p. 301; Ruse, "Philosopher's Day", p. 26. As Ruse puts it: "Even if Scientific Creationism were totally successful in making its case as science, it would not yield a 'scientific' explanation of origins. The Creationists believe that the world started miraculously. But miracles lie outside of science, which by definition deals only with the natural, the repeatable, that which is governed by law" (*Darwinism*, p. 182). Richard Lewontin expresses a similar fear in *Scientists Confront Creationism*: "Either the world of phenomena is a consequence of the regular operation of repeatable causes and their repeatable effects, operating roughly along the lines of known physical law, or else at every instant all physical regularities may be ruptured and a totally unforeseeable set of events may occur. . . . We can not live simultaneously in a world of natural causation and of miracles, for if one miracle can occur, there is no limit" ([New York: Norton, 1983], p. xxvi).

[61] This dichotomy between "unbroken law" and the action of agency is merely a species of the same genus of confusion that led Ruse and others to insist that science always explains via laws. In Ruse's case the dichotomy is manifest in his assertion that invoking the action of a divine agent constitutes a "violation of natural law". I disagree. Pitting the action of agents (whether seen or unseen) against natural law creates a false opposition. The reason for this is simple. Agents can change initial and boundary conditions, yet in so doing they do not violate laws. Most scientific laws have the form "If A, then B will follow, given conditions X." If X is altered or if A did not obtain, then it constitutes no violation of the laws of nature to say that B did not occur, even

if we expected it to. Agents may alter the course of events or produce novel events that contradict our expectations without violating the laws of nature. To assert otherwise is merely to misunderstand the distinction between antecedent conditions and laws. C. S. Lewis, *God in the Dock* (London: Collins, 1979), pp. 51–55. See R. Swinburne, *The Concept of Miracle* (London: Macmillan, 1970), pp. 23–32, and G. Colwell, "On Defining Away the Miraculous", *Philosophy* 57 (1982): 327–37, for other defenses of the possibility of miracles that assume and respect the integrity of natural laws.

[62] See also Kavalovski, *Vera Causa*, pp. 104–29, for a discussion of the so-called *vera causa* principle, a nineteenth-century methodological principle invoked by Darwin to eliminate from consideration creationist explanations judged to be unobservable (Darwin, *Origin*, pp. 201, 430, 453).

[63] Skoog, "View"; Gould, "Genesis", pp. 129–30; Ruse, "Witness", p. 305.

[64] Grinnell, "Radical Intersubjectivity: Why Naturalism Is an Assumption Necessary for Doing Science", paper presented at the conference on "Darwinism: Scientific Inference or Philosophical Preference?" Southern Methodist University, Dallas, March 26–28, 1993.

[65] Skoog, "View".

[66] S. C. Meyer, "A Scopes Trial for the '90s", *The Wall Street Journal*, December 6, 1993, p. A14; S. C. Meyer, "Open Debate on Life's Origin", *Insight*, February 21, 1994, pp. 27–29. Eugenie Scott, "Keep Science Free from Creationism", *Insight*, February 21, 1994, p. 30.

[67] H. Judson, *The Eighth Day of Creation* (New York: Simon and Schuster, 1979), pp. 157–90.

[68] Meyer, *Of Clues*, p. 120; Darwin, *Origin*, p. 398; D. Hull, *Darwin and His Critics* (Chicago: University of Chicago Press, 1973), p. 45.

[69] C. Darwin, *More Letters of Charles Darwin*, ed. F. Darwin, 2 vols. (London: Murray, 1903), 1:184.

[70] Quoted in S. J. Gould, "Darwinism Defined: The Difference between Theory and Fact", *Discovery*, January 1987, p. 70.

[71] Darwin's use of both methodological and empirical arguments against creationism has been well documented: Gillespie, *Darwin*, pp. 67–81; Kavalovski, *Vera Causa*, pp. 104–29; Meyer, *Of Clues*, pp. 123–25; Recker, "Efficacy", p. 173; Hull, "Darwin", pp. 63–80. For examples of Darwin's methodological arguments, see Darwin, *Origin*, pp. 201, 430, 453. For examples of his empirical arguments, see *Origin*, pp. 223, 386, 417–18.

[72] Skoog, "View".

[73] Kitcher, *Abusing Science*, p. 125. While Kitcher allows for the possibility of a testable theory of divine creation, he believes creationism was tested and found wanting in the nineteenth century.

[74] I am currently undertaking an exhaustive cataloguing and evaluation of evolutionary demarcation arguments. Those arguments not discussed here will

be addressed in subsequent work published through the Pascal Centre in Ontario, Canada.

[75] This phrase is actually used by astronomer Carl Sagan (in Carl Sagan and Ann Druyan, *Shadows of Forgotten Ancestors* [New York: Random House, 1992], p. 387) but clearly expresses the posture of many evolutionary gradualists and punctuationalists with respect to the absence of transitional intermediates in the fossil record.

[76] The same could be said of the neo-Darwinian selection-mutation mechanism vis-à-vis the theory of common descent. In both cases, however, issues of warrant and issues of scientific status should not be confused.

[77] For a design argument not based on religious authority (i.e., contra g: "Creationist or design theories are not tentative"), see M. Denton, *Evolution: A Theory in Crisis* (London: Adler and Adler, 1986), pp. 338–42. For an examination and refutation of demarcation argument h (i.e., "Creationist or design theories have no problem-solving capability"), see J. P. Moreland's forthcoming "Scientific Creationism, Science and Conceptual Problems", in *Perspectives on Science and Christian Faith*.

[78] C. S. Peirce, "Abduction and Induction", in *The Philosophy of Peirce*, ed. J. Buchler (London: Routledge, 1956), pp. 150–56; C. S. Peirce, *Collected Papers*, ed. C. Hartshorne and P. Weiss, 6 vols. (Cambridge: Harvard University Press, 1931), 2:375; K. T. Fann, *Peirce's Theory of Abduction* (The Hague: Martinus Nijhoff, 1970), p. 33; Meyer, *Of Clues*, pp. 24–34.

[79] Meyer, *Of Clues*.

[80] These three features can be used as a set of individually necessary and jointly sufficient conditions for the identification of historical, as opposed to nonhistorical, sciences. Nevertheless, this demarcation or definition is admittedly arbitrary. It does not imply that some sciences do not combine elements of both historical and inductive inquiry, or that many disciplines do not have both inductive and nomological branches—e.g., cosmology and cosmogony. This "demarcation" is also unproblematic because it makes no claim, implicit or explicit, for a privileged epistemological status for disciplines that manifest historical features. The distinction is not, however, without justification, since each of the individually necessary conditions of a historical science do distinguish real qualitative or logical differences between types of inferences, explanations, or questions.

[81] A. C. Doyle, "The Boscome Valley Mystery", in *The Sign of Three: Peirce, Holmes, Popper*, ed. T. Sebeok (Bloomington: Indiana University Press, 1983), p. 145.

[82] S. J. Gould, "Evolution and the Triumph of Homology: Or, Why History Matters", *American Scientist* 74 (1986): 61.

[83] This is not to deny that laws or process theories may play roles in support of causal explanation, as even opponents of the covering-law model such as Scriven admit. Scriven notes that laws and other types of general process

theories may play an important role in justifying the causal status of an explanatory antecedent and may provide the means of inferring plausible causal antecedents from observed consequents. Nevertheless, as both Scriven and I have argued elsewhere, laws are not necessary to the explanation of particular events or facts; and even when laws are present, antecedent *events* function as the primary causal or explanatory entity in historical explanations. Scriven, "Truisms", pp. 448–50; Scriven, "Explanation", p. 480; Scriven, "Causes", pp. 249–50; Meyer, *Of Clues*, pp. 18–24, 36–72, 84–92.

[84] Meyer, *Of Clues*, pp. 112–36.

[85] C. Darwin, *The Descent of Man*, 2d ed. (London: A. L. Burt, 1874), p. 61.

[86] Darwin, *Origin*, p. 434. Darwin's next line on the following page and the very first line of his concluding chapter again suggest the primacy of his concern to establish "descent with modification" and the supportive role that natural selection played in his argument. In his words: "As this whole volume is one long argument, it may be convenient to the reader to have the leading facts and inferences briefly recapitulated. That many grave objections may be advanced against [a] the theory of descent with modification [b] through natural selection, I do not deny."

[87] Ibid., pp. 331–434.

[88] Ibid., p. 432.

[89] Ibid., p. 434.

[90] Gould, "Evolution", p. 61.

[91] Quoted in Gould, "Darwinism", p. 70.

[92] Darwin, *Origin*, p. 399.

[93] Ibid., pp. 195, 399. As Kavalovski has noted, Darwin did not limit his claim of *vera causa* to natural selection but included descent itself under this appellation (Kavalovski, *Vera Causa*, pp. 104–5). In chapter 5, on "Laws of Variation", Darwin refers explicitly to "community of descent" as a *vera causa* of homologies among plant species (*Origin*, p. 195). Despite many references to natural selection as a *vera causa* of morphological change in general, Darwin also seemed to recognize the need to postulate a historical cause (i.e., a pattern of past events) to explain the particular facts mentioned above. Darwin makes this relationship between causal postulations about the past and explanations of present phenomena explicit at one point in chapter 13 by stating that "we may thus account even for the distinctness of whole classes from each other . . . by the belief that many ancient forms of life *have been utterly lost*" (*Origin*, p. 413).

[94] Gould, "Evolution", p. 60.

[95] A. I. Oparin, *The Origin of Life*, trans. S. Morgulis (New York: Macmillan, 1938).

[96] Meyer, *Of Clues*, pp. 237–40.

[97] Quoted in Gould, "Darwinism", p. 70.

⁹⁸ F. Darwin, ed., *Life and Letters of Charles Darwin*, 2 vols. (London: D. Appleton, 1896), 1:437.

⁹⁹ Denton, *Evolution*, pp. 338–42; C. Thaxton, W. Bradley, and R. Olsen, *The Mystery of Life's Origin* (New York: Philosophical Library, 1984), pp. 113–65, 201–4, 209–12.

¹⁰⁰ Thaxton, Bradley, and Olsen, *Mystery*, pp. 201–12.

¹⁰¹ E. J. Ambrose, *The Nature and Origin of the Biological World* (New York: Halstead, 1982); Denton, *Evolution*; R. Augros and G. Stanciu, *The New Biology* (Boston: Shambhala, 1987); D. Kenyon and P. W. Davis, *Of Pandas and People: The Central Question of Biological Origins* (Dallas: Haughton, 1993).

¹⁰² D. Hull, "God of the Galápagos", *Nature* 352 (1991): 485–86.

¹⁰³ Such a concern was recently raised, for example, in Nancey Murphy's critique of Phillip Johnson's book *Darwin on Trial* (N. Murphy, "Phillip Johnson on Trial: A Critique of His Critique of Darwin", *Perspectives on Science and Christian Faith* 45, no. 1 [1993]:34). There Murphy cites concern among theistic scientists about the God-of-the-gaps objection as a reason for the exclusion of creative intelligence as a candidate explanation for the origin of life. As Murphy explains, even many theistic scientists worry that theistic explanations give up on science too soon, thus making the God hypothesis vulnerable to future scientific advance. Yet clearly these scientists accept a definition of science and scientific advance that presupposes the very naturalism already asserted as necessary to science. Why can a theistic explanation not constitute a scientific advance? Murphy offers no answer to this question, beyond her reference to the story of Laplace's mathematical model supplanting Newton's interventionist explanation of planetary motion.

¹⁰⁴ See note 36.

¹⁰⁵ Meyer, *Of Clues*, pp. 77–136.

¹⁰⁶ Thaxton, Bradley, and Olsen, *Mystery*, pp. 113–65, 201–4, 209–12; Kenyon and Davis, *Pandas*; W. Bradley and C. B. Thaxton, "Information and the Origin of Life", in *The Creation Hypothesis: Scientific Evidence for an Intelligent Designer*, ed. J. P. Moreland (Downers Grove, Ill.: InterVarsity Press, 1994), pp. 173–210.

¹⁰⁷ W. A. Dembski, "Redesigning Science", in *Mere Creation: Science, Faith and Intelligent Design*, ed. W. A. Dembski (Downers Grove, Ill.: InterVarsity Press, 1998), pp. 93–112; W. A. Dembski, *Intelligent Design* (Downers Grove, Ill.: InterVarsity Press, 1999), pp. 122–52.

¹⁰⁸ M. Behe, *Darwin's Black Box* (New York: Free Press, 1996).

¹⁰⁹ Note that I do not need to know something about my neighbor's proclivities, character, or purposes (in this case, his dislike of yard debris) in order to make a secure inference to intelligent design, though knowing something about his proclivities does strengthen my confidence in the inference that I have drawn. I can know *that* something was designed without knowing *why* or *who* designed it.

[110] Dembski, "Redesigning Science", pp. 93–112; Dembski, *Intelligent Design*, pp. 127–46.

[111] Murphy, "Johnson", p. 33.

[112] As Newton wrote to Bentley in 1692: "It is inconceivable that inanimate brute Matter should, without the Mediation of something else which is not material, operate upon and affect other Matter without mutual Contact" (Newton, *Papers*, p. 302).

[113] In any case, none of the emphasis on the regularity and constancy of laws prevented either Boyle or Newton from invoking special divine action as an explanation for the origin of particular natural features (M. A. Stewart, ed., *Selected Philosophical Papers of Robert Boyle* [New York and Manchester: Manchester University Press, 1979], p. 144). Boyle postulated design for the origin of atomic structure; Newton did so in optics and astronomy. Those (such as Murphy, "Johnson", p. 33) who cite these two men as the source of the current positivistic prohibition against mixing science and metaphysics are simply incorrect. (Instead they should consult Gillespie, *Darwin*, pp. 1–66 on Darwin's positivism.) Boyle in fact devised an interesting classification scheme that makes explicit the metaphysical nonneutrality of origins questions (Stewart, *Boyle*, pp. 172–75), which he thought occurred in a region where natural philosophy and religion overlapped. While Newton tended to reserve the term natural philosophy for nomological disciplines, he in no sense agreed that empirical evidence was metaphysically neutral, for the reasons already stated.

[114] C. B. Kaiser, *Creation and the History of Science*, History of Christian Theology Series, vol. 3 (Grand Rapids, Mich.: William B. Eerdmans Publishing Company, 1991), p. 264.

[115] Following Sober, I regard simplicity as a notion that cannot be formally explicated but which, nevertheless, plays a role in scientific theory evaluation. Like Sober I believe that intuitive notions of simplicity, economy, or elegance express or are informed by tacit background assumptions. I see no reason that theistic explanations could not be either commended or disqualified on the basis of such judgments just as materialistic explanations are. Sober, *Reconstructing*, pp. 36–69.

[116] Theists who invoke the special assistance or activity of divine agency to explain an origin event or biblical miracle, for example, are not, as is commonly asserted, guilty of semideism. Those who infer that God has acted in a discrete, special, and perhaps more easily discernible way in one case do not deny that he is constantly acting to "uphold the universe by the word of his power" at all other times. The medievals resisted this false dichotomy by affirming two powers of God, or two ways by which he interacts with the world. The ordinary power of God they called his *potentia ordinata* and the special or fiat power they called his *potentia absoluta*. W. Courtenay, "The Dialectic of Omnipotence in the High and Late Middle Ages", in *Divine Omniscience and Omnipotence in Medieval Philosophy*, ed. T. Rudavsky (Norwell: Kluwer Academic Publishers,

1984), pp. 243–69. Many modern theists who affirm the special action of God at a discrete point in history have this type of distinction in mind.

[117] It may sound as though I am endorsing a philosophical relativism about science, or the kind of methodological anarchism advocated by the philosopher of science Paul Feyerabend in his book *Against Method* (London: Verso, 1978). Quite the contrary: I am not an antirealist, nor do I deny the importance of methodology to the process of formulating warranted belief. Precisely because I recognize the importance of a great number of quite distinct and well-established methods at work within fields already widely acknowledged to be scientific, I deny the utility of attempts to give *a* single, universal methodological characterization of science.

[118] For example, theories that offer antecedent conditions that are merely necessary to a given outcome do not succeed logically as explanations of that outcome. The methodological convention extant within most historical sciences requiring postulated antecedents to meet a criterion of etiological plausibility (causal adequacy) expresses this logical requirement. See my discussion of the logical and contextual requirements of causal explanation in Meyer, *Of Clues*, pp. 60–71, 84–92.

[119] The logical and epistemic conditions of successful causal explanation are difficult to make explicit, though they are quite easy to apply apparently via a kind of tacit understanding. For a more comprehensive (explicit) discussion of the logical and contextual requirements of causal explanation, see ibid., pp. 36–76.

[120] G. K. Chesterton, *Orthodoxy* (London: John Lane, 1909).

[121] As Basil Willey put it: "Science must be provisionally atheistic or cease to be itself" ("Darwin's Place", p. 15). See also Ruse, *Darwinism*, p. 59; Ruse, "Witness", p. 305; Gould, "Evolution", p. 121; Root-Bernstein, "Creationism Considered", p. 74; Ruse, "Darwinism", pp. 1–13.

[122] W. A. Dembski, "The Very Possibility of Intelligent Design", paper presented at "Science and Belief", the first international conference of the Pascal Centre, Ancaster, Ontario, August 11–15, 1992.

[123] P. Thagard, "The Best Explanation: Criteria for Theory Choice", *Journal of Philosophy* 75 (1978): 79; Meyer, *Of Clues*, pp. 99–109; W. Whewell, *The Philosophy of the Inductive Sciences*, 2 vols. (London: Parker, 1840), 2:109, 242; L. Laudan, "William Whewell on the Consilience of Induction", *The Monist* 55 (1971): 371–79.

[124] Lipton, *Inference*.

[125] See Haeckel, *Wonders*, pp. 110–11.

[126] Johnson, *Darwin*. See also Gillespie, *Darwin*, pp. 1–18, 41–66, 146–56, for an interesting discussion of the way Darwin succeeded in redefining science so as to make creationist or idealist dissent impossible from within science.

WILLIAM A. DEMBSKI
AND STEPHEN C. MEYER

FRUITFUL INTERCHANGE OR POLITE CHITCHAT? THE DIALOGUE BETWEEN SCIENCE AND THEOLOGY

In his intellectual autobiography Rudolf Carnap observed, "If one is interested in the relations between fields which, according to customary academic divisions, belong to different departments, then one will not be welcomed as a builder of bridges, as one might have expected, but will rather be regarded by both sides as an outsider and troublesome intruder."[1] Carnap learned the hazards of interdisciplinary bridgebuilding by bitter experience. To this day philosophers recall how Carnap's efforts to relate philosophy and physics were obstructed during his stint at the University of Chicago's philosophy department in the 1940s and 1950s.

Since Carnap's day, and in part because of Carnap's efforts, the bridges between philosophy and physics have become more firmly established, with philosophy of science and, in particular, philosophy of physics now accepted as legitimate subdisciplines of philosophy. Moreover, certain philosophers of physics have through their work gained recognition in the physics community. Those who come to mind include Abner Shimony, who holds doctorates in both physics and philosophy; David Mala-

This essay first appeared in *Zygon* 33, no. 3 (September 1998): 415–30, and is reprinted by permission of Blackwell Publishers.

The authors wish to thank Professor Wentzel van Huyssteen of Princeton Theological Seminary for his valuable insights and suggestions.

ment, who has proved technical mathematical results in general relativity; and Arthur Fine, who has done original work on the foundations of quantum mechanics.

Still, it would be an overstatement to say that philosophers and physicists are engaged in active dialogue. Philosophy has traditionally been classified with the humanities, and physics with the natural sciences. Much of what philosophers do and much of what physicists do simply do not intersect. A moral philosopher's metaethical ruminations on the nature of duty and a physicist's tinkering with lasers in the laboratory do not seem to connect in relevant ways. Given this perception, it is not hard to see why interested outsiders are often regarded as pesky meddlers, not as individuals from whom disciplinary insiders might actually learn something pertinent to their endeavors.

The difficulties attendant on the interdisciplinary conversation between physics and philosophy, and between the humanities and the natural sciences more generally, often pale by comparison to those encountered in the interdisciplinary dialogue between theology and the natural sciences. Distinct disciplines have a hard time communicating, even those which prima facie we might think would want to communicate, for example, philosophy and physics. How much more difficult it is, then, to get theology and science communicating when, especially over the last one hundred years, they have been increasingly characterized in terms of either a warfare or a partition metaphor (that is, either they are in unresolvable conflict or they are so thoroughly compartmentalized that no possibility of meaningful communication exists).

But let us suppose for the sake of argument that we are in a world, not of ideal rational agents, but of ideal amicable agents —amicable in the sense that the agents are willing to talk to, listen to, and learn from each other. In such a world, would a dialogue between theology and science be fruitful? Would it further inquiry? Would it foster an increased understanding of the world? Would it yield a net gain of knowledge to both theologian and scientist? Or would only one side in the dialogue profit? Would such a dialogue constitute merely polite chitchat between members of different intellectual communities? Would they at the end of the day conclude that nothing of any genuine consequence had been accomplished through the dialogue?

On the assumption that we in the sciences and theology are willing to communicate with and listen attentively to each other, let us pose the question: Are there any good reasons to think that scientists and theologians will actually learn something from each other's disciplines that will be valuable to their own? To be sure, both will learn some new things from such a dialogue. The theologian may learn from the physicist that the universe began as an incredibly dense fireball whose genesis is known as the Big Bang, whereas the physicist may learn that theologians' believe that God created the world by means of a divine logos. So the theologian and the physicist will each have a new piece of information about another discipline to add to their stock of knowledge. But how are these pieces of information to be integrated into the web of information that constitutes our knowledge of the world? And how might a theological piece of information affect a physicist's physical understanding of the world, and *mutatis mutandis*, how might a piece of information from physics affect a theologian's theological understanding of the world?

What underlies these questions is the issue of *epistemic support*. In the context of an interdisciplinary dialogue, epistemic support is concerned with how the acceptance of claims in one discipline might justify the acceptance of claims in another. Now philosophers have written extensively about epistemic support and say that their work is directly relevant to evaluating the nature of the dialogue between theology and science. Nevertheless, if we are naive in how we appropriate their work, we will come to an impasse in the interdisciplinary dialogue between theology and science. This essay will therefore seek to describe a conception of epistemic support that fosters a genuinely productive interdisciplinary dialogue between theology and science.

How, then, should we characterize the form epistemic support takes in the dialogue between theology and science? What will it mean for a scientific (alternatively, theological) claim A to support a theological (alternatively, scientific) claim B? Does it mean that B follows as a logical deduction from A, or that there is an airtight circumstantial case to be made for B, given A, or that it is irrational to reject B once A is taken for granted? Support in any of these senses is the very strong notion of rational compulsion. The notion of support for which we argue in this essay is considerably weaker and will be explicated by reference to explanatory power.

Failure to distinguish between a strong and a weak form of epistemic support has led to confusion in the dialogue between science and theology. Consider, for instance, what Ernan McMullin means when he denies that the relation between the Big Bang and the creation of the universe by God can be characterized in terms of epistemic support: "What one could say . . . is that if the universe began in time through the act of a Creator, from our vantage point it would look something like the Big Bang that cosmologists are talking about. What one cannot say is, first, that the Christian doctrine of Creation 'supports' the Big Bang model, or, second, that the Big Bang model 'supports' the Christian doctrine of Creation."[2] Contra McMullin, we insist that the Big Bang model does support the Christian doctrine of Creation, and vice versa. Yet we will develop a more liberalized notion of epistemic support that allows fruitful interdisciplinary dialogue without requiring that scientific evidence compel religious beliefs or the reverse.

Rational Compulsion

Rational compulsion involves a far stronger notion of support than typically comes up within either science or theology, much less in the dialogue between the two. One feels rationally compelled to believe necessary truths like $2 + 2 = 4$. One may even feel rationally compelled to believe in the existence of certain medium-sized objects such as trees, cars, and people.[3] Nevertheless, a considerably weaker conception of epistemic support seems to prevail in science and theology and seems appropriate for characterizing any interdisciplinary connections between the two.

Our primary task, then, is to delineate a conception of epistemic support whereby the interdisciplinary dialogue between science and theology does not reduce to idle chitchat but can instead engender deeper understanding and sponsor further inquiry. Recent developments in the philosophy of science make possible just such a conception of epistemic support.[4] Nevertheless, before describing these, we wish to indicate by way of negation the form epistemic support must not take if it is to foster genuinely productive interdisciplinary dialogue between theology and science.

The *bête noire* that has at every turn obstructed meaningful dialogue between theology and science is the demand that epistemic support be conceived as some form of rational compulsion. Rational compulsion is our own term, but it seems to capture the conception of epistemic support presupposed in so many ill-fated attempts to bring science and theology into dialogue. It may therefore be helpful to consider this conception of epistemic support in some detail. First off, let us specify that rational compulsion constitutes a perfectly valid form of epistemic support. Indeed, if A rationally compels B, then it is irrational to deny B if one affirms A. In such cases, A clearly provides epistemic support for B.

In practice rational compulsion takes the form of an entailment relation, either strict or partial. For A strictly to entail B means that it is impossible for A to be true but B false. Strict entailment is typically what people mean when they refer to deduction or demonstration or proof. On the other hand, for A partially to entail B means that the conditional probability of B given A is greater than the unconditional probability of B by itself. Partial entailment is equivalent to what goes by the technical name probabilification, though partial entailment is not coextensive with the more classical notion of probable reasoning. Partial entailment is a more general notion than strict entailment and properly subsumes it, because A strictly entails B just in case the conditional probability of B given A is one.[5]

Whether strict or partial, entailment is a logical relation, with the directionality of the logic going from the antecedent to the consequent (that is, from the thing doing the entailing to the thing entailed). In practice we know that A strictly entails B when we can find a logical argument that takes A as a premise and which by a series of logical machinations (usually deductions according to certain inference rules) leads to B as a consequence. On the other hand, we know that A partially entails B when we have reliable ways of assigning probabilities to claims involving A and B and find that the conditional probability of B given A is greater than the unconditional probability of B by itself.

We wish to stress that both strict and partial entailment yield what we have been calling rational compulsion. This is immediately obvious for strict entailment. Indeed, if it is impossible for B to be false if A is true, then if we affirm A we surely had better affirm B also. Still, we may wonder why partial entailment should also yield rational compulsion. Whereas strict entailment leaves no room for either (1) fallibility or (2) contingency or (3) degree or (4) doubt, partial entailment leaves room for all of these. If A strictly entails B, then (1) there is no possibility of being wrong about B if we are right about A; (2) B follows necessarily from A; (3) A epistemically supports B to the utmost and cannot be made to support B to a still higher degree; and (4) not only need we not but we also ought not to doubt B if we trust A.

On the other hand, none of these properties holds in general for partial entailment. Consider the following two claims:

A: There will be a heavy snowfall tonight.
B: Schools will be closed tomorrow.

Suppose that nine times out of ten when there is a heavy snowfall at night, schools close on the next day. Then if we see heavy snow accumulating tonight, we have good reason to expect that school will be closed tomorrow. Nevertheless, the four claims we just made about strict entailment in the last paragraph fail to hold for partial entailment. Thus (1) even though A may hold, we may still be mistaken for holding B; (2) there is no necessary connection between A and B; (3) the relation of support between A and B admits of degrees (for instance, the relation would be still stronger if ninety-nine times out of a hundred school were closed following a heavy snowfall, weaker if only two times out of three); and (4) we are entitled to invest B with a measure of doubt even if we know A to be true.

Nevertheless, partial entailment and rational compulsion remain inextricably linked. To see this, consider the following rumination by C. S. Peirce:

> If a man had to choose between drawing a card from a pack containing twenty-five red cards and a black one, or from a pack containing twenty-five black cards and a red one, and if the drawing of a red card were destined to transport him to eternal felicity, and that of a black one to consign him to everlasting woe, it would be folly to deny that he ought to prefer the pack containing the larger portion of red cards, although, from the nature of the risk, it could not be repeated. . . . But suppose he should choose the red pack, and should draw the wrong card, what consolation would he have?[6]

Yes, you might end up with a black card if you choose from the deck consisting predominantly of red cards, but you will be much more likely to end up with a black card if you choose from the other deck. Hence, if your aim is to avoid everlasting woe, you had better choose a card from the predominantly red deck. Now the injunction "you had better choose the red deck" is certainly a form of rational compulsion.

Thus, rational compulsion arises even when we are dealing, not with certainties, but with probabilities. Suppose therefore that A and B are claims and that P is a probability that handles claims involving A and B. Then if P(B|A) (the conditional probability of B given A) is greater than P(B) (the unconditional probability of B), we are rationally compelled or obligated to invest more credence in B on the assumption of A than in B taken by itself. Moreover, since it is a basic property of probabilities that P(B|A) = 1 − P(~B|A) (~B is the negation of B), it follows that whenever P(B|A) is greater than 1/2, then P(~B|A) is less than 1/2. Thus, if we know that A has happened and that P(B|A) is greater than 1/2, then if we must base a course of action on whether B occurs or not, we must suppose that B, and not its negation, will occur.[7] In this way we see that not only strict entailment but also partial entailment yields a form of rational compulsion.

The question now remains, Why will not rational compulsion do as an account of epistemic support in the dialogue between science and theology? We see two problems with this standard.

First, the logic of entailment constitutes an excessively restrictive conception of epistemic support for science itself. Scientists can rarely prove their theories from empirical evidence in either of the two senses of entailment discussed above. Indeed, no field of inquiry short of mathematics could progress if it limited itself to the logic of strict or partial entailment. Rather, most fields of inquiry employ alternate forms of inference known variously as the method of hypothesis, the hypothetico-deductive method, abduction, or inference to the best explanation. Yet the limits inherent in the logic of both scientific prediction and explanation ensure that even good theories cannot be affirmed with certainty without also committing the fallacy of affirming the consequent. In the language of contemporary philosophy of science, empirical data often leave scientific theories underdetermined. Yet if scientists cannot prove (or make rationally compelling) their own theories from empirical data, it seems doubtful that theologians will succeed in proving theological doctrines from data in the same way. Thus, it follows that if rational compulsion stands as the only way for science to provide epistemic support for theology, little fruitful dialogue between the two disciplines will occur. Indeed, since empirical evidence rarely compels (in the sense defined above) acceptance of theories within science, it seems likely that the demand for rational compulsion will generally stultify interdisciplinary dialogue between theology and science.

Yet rational compulsion creates another impediment to productive interdisciplinary dialogue. In the logic of entailment, logic and epistemic support move in the same direction. If A rationally compels B, then A strictly or partially entails B *and* A epistemically supports B. For a relation of epistemic support between A and B to obtain, the thing that does the supporting, in this case A, must be taken for granted—A must be given. But once A is given, any consequences strictly or partially entailed by A, say B, must be accepted as well—after all, A rationally compels B.

This creates a problem for interdisciplinary dialogue because presumably it is the implications of evidence from a given field that interest, for example, a theologian. Yet because the logic of entailment makes it irrational for anyone to doubt B given A, the theologian must either accept the implications of the scientific data without further discussion or challenge the evidential premise for the entailed conclusion (which the theologian, as a nonscientist, is in no position to do). Suppose, for example, that B follows from some evidential claim A from within a scientific discipline. And suppose, as is sometimes, though rarely, the case, that A happens strictly or partially to entail B. Suppose further that scientists generally are firmly committed to B, but the theologian finds B repugnant. For instance, we might imagine a dialogue between a member of the scientific establishment and a biblical scholar or theologian committed to a young earth. In this case, A is the claim that radiometric dating methods are sound, and B the claim that the earth is several billion years old. Here granting A does strictly entail B. But since the theologian is committed to an earth that is only a few thousand years old, B is utterly unacceptable. What then does the theologian do? The standard practice of the biblical scholar is to impugn A, that is, to reject the radiometric dating methods. Thus, the interdisciplinary dialogue between the biblical scholar or theologian and the scientific establishment does not even get off the ground. What is a fundamental assumption for the scientist, namely, A, entails a conclusion that is unacceptable to the theologian.

Of course, many theologians may adopt a less combative posture relative to scientific evidence or theory. Yet if the possible epistemic importance for theology of some evidence or theory A is confined to its logically entailed consequences B, even more scientifically sympathetic theologians may find little to contribute to an interdisciplinary dialogue—if for no other reason than it is irrational to doubt B given the logical exigencies of entailment. If, for the theologians, nothing is riding on some proposition B, then the theologians can graciously accept B, if A happens to entail B. Yet in this instance, the theologians do not learn anything genuinely new or significant about their discipline, nor do they contribute to understanding the science represented by A. In this case, B is irrelevant, or at best oblique, to the theologians' concerns; in the other, B so utterly contradicts the theologians' beliefs as to create irreconcilable conflict. Yet in neither case does fruitful dialogue ensue. Instead, presupposing that only the logic of entailment is relevant to the science and theology dialogue creates a conversation often characterized by either hostile accusation or polite chitchat.

Explanatory Power[8]

We believe an alternative understanding of epistemic support can foster a more productive interdisciplinary dialogue between science and theology. Fortunately such an alternative under-standing is available. Although there are a number of ways to approach this alternative understanding of epistemic support, we approach it through the notion of *explanatory power*.[9]

A little history will help clarify what we mean by explanatory power. During the last century, C. S. Peirce devoted consider-able energies to describing the modes of inference by which we derive conclusions from data. Because data are given and conclusions depend for their justification upon data, the rela-tion of epistemic support is invariably directed from data to conclusion. Thus, if A comprises the data and B the conclu-sion, we say that A provides evidence for, serves to confirm, or epistemically supports B (where each of these expressions amounts to the same thing).

Now the thing Peirce observed is that the direction of the logic relating A and B need not go in the same direction as the relation of epistemic support between A and B. In the case of rational compulsion and entailment, as we saw in the last sec-tion, the directions are identical. Nevertheless, it can happen that the relation of epistemic support goes in one direction but the logic relating data and conclusion goes in the other. Peirce used the term *deduction* to characterize inference patterns whose logic and support relations were directed similarly, whereas he used the term *abduction* to characterize those where they were directed oppositely.[10]

The difference between these inference patterns becomes ap-parent from the following argument schemata:[11]

Deduction Schema

DATA: A is given and plainly true.
LOGIC: But if A is true, then B is a matter of course.

CONCLUSION: Hence, B must be true as well.

Abduction Schema

DATA: The surprising fact A is observed.

LOGIC: But if B were true, then A would be a matter of course.

CONCLUSION: Hence, there is reason to suspect that B is true.

Notice that the data and the conclusion of both schemata are identical, for in both instances we are given A and we conclude B. Yet the logic is entirely reversed. In the deduction schema the logic proceeds from A to B, whereas in the abduction schema the logic proceeds from B to A.

The logic of the deduction schema is the logic of entailment. Once A is given, anything logically entailed by A must be accepted as well. Within the deduction schema valid conclusions are therefore those entailed by A. The logic of the abduction schema, on the other hand, hinges on a quite different logic, one we shall call the *logic of explanation*. Once A is given, anything that neatly explains A becomes highly plausible. Within the abduction schema valid conclusions are therefore those that explain A.

Now it needs to be stressed that the logic of explanation is incompatible with the logic of deduction. As far as the logic of deduction is concerned, the logic of explanation commits the fallacy of affirming the consequent. The fallacy of affirming the consequent is essentially a failure to acknowledge that antecedent conditions can be underdetermined, that is, that more than one antecedent might explain the same evidence.

For instance, suppose we know that Frank was promoted, and suppose we know that if Frank behaves obsequiously toward his boss, he will be sure to be promoted. It does not follow as a logical deduction that Frank did in fact behave obsequiously toward his boss. Frank may just be incredibly competent so that his boss decided to promote him despite his not being obsequious. Alternatively, Frank's mother may be the head of the company, and so Frank's boss thought it wise to promote Frank even though Frank was at times downright rude. The point is that the explanation of Frank's promotion (whether it was on account of his obsequious behavior or on account of his mother being company president or whatever) is not governed by the logic of deduction. In particular, the logic of explanation involves no rational compulsion.

Peirce admitted as much when he noted, "As a general rule [abduction] is a weak kind of argument. It often inclines our judgment so slightly toward its conclusion that we cannot say that we believe the latter to be true; we only surmise that it may be so."[12] Yet as a practical matter Peirce acknowledged that abduction often yields conclusions that are difficult to doubt even if they lack the airtight certainty that accompanies the logic of deduction. For instance, Peirce argued that skepticism about the existence of Napoleon Bonaparte was unjustified even though Napoleon's existence could be known only by abduction. As Peirce put it, "Numberless documents refer to a conqueror called Napoleon Bonaparte. Though we have not seen the man, yet we cannot explain what we have seen, namely, all these documents and monuments, without supposing that he really existed." To this Peirce added, "There is no difference except one of degree between such an [historical] inference and that by which we are led to believe that we remember the occurrences of yesterday from our feelings as if we did so."[13]

To sum up, whereas in the logic of deduction, A epistemically supports B because A logically entails and therefore rationally compels B, in the logic of explanation, A epistemically supports B because B provides a good explanation of A. As Peirce showed, both logics provide legitimate inference patterns and underwrite robust relations of epistemic support. Yet although these logics often work in tandem, they are nevertheless distinct. Moreover, the logic of explanation suggests an important role for theology in enhancing our understanding of some scientific data, results, or theories. Unlike the logic of entailment, which left theology little to do beyond (in the most negative case) questioning the empirical findings of science, the logic of explanation suggests that theology might provide science with a source of (albeit in many cases metaphysical) hypotheses and explanations for its empirical findings and results. This logic further suggests a way that scientific data might provide epistemic support for theological propositions or doctrines. In particular, it suggests that scientific data can provide epistemic support for theological propositions just in case those propositions suggest a better explanation for the data than do the alternatives under consideration.

Contemporary Developments

What has happened to the logic of explanation and its concomitant conception of epistemic support since Peirce's day? The key development has been a generalization of Peircean abduction via the notion of explanatory power. Even though Peirce clearly distinguished deduction from abduction, there is a sense in which deduction still plays a central role within Peircean abduction. Recall the Peircean abduction schema:

Abduction Schema

DATA: The surprising fact A is observed.
LOGIC: But if B were true, then A would be a matter of course.

CONCLUSION: Hence, there is reason to suspect that B is true.

Within the logic of this abduction schema, the prototypical example of B explaining A is the case in which A follows as a logical deduction from B (alternatively, B strictly entails or rationally compels A). Thus, as an elementary example of abduction, Peirce considered the case where A = *every bean observed from the bag is white* and B = *all the beans in the bag are white*.[14] Here B not only explains A but actually entails A (indeed, there is a one-step logical deduction leading from B to A).

Of course, Peirce also understood that such strict entailment relations were not necessary to provide an explanation. Nevertheless, he gave no account of how a rational agent might assess which of the many possible (abductively inferred) hypotheses might stand as the best explanation of some evidence A. In recent years, however, philosophers of science have clarified how such assessments are made. They have proposed three criteria that must be satisfied in order for B to constitute the best explanation of A. These are as follows:

First, B must be *consonant* with A.[15] Thus, instead of injecting discord or dissonance into our understanding of A, B must harmonize with A as well as the network of beliefs of which A is a part. In particular, if one were to take B as an (abductive) hypothesis, one would expect A to follow as a matter of course. To say that B is consonant with A implies that A confirms B, where B is taken as a hypothesis. Note that consonance is more than simply a coherentist requirement. Consonance involves both goodness of fit and aesthetic or theoretical judgment. A and B must not only be at peace with each other but one ought to have some reason to expect A given B.[16]

Second, B must *contribute* to A. Thus, B must perform some useful work in helping to explain A. B must solve problems or answer questions pertinent to A which could not be handled otherwise. This second requirement is a corollary of Occam's razor, ensuring that adding B to our stock of beliefs will not be superfluous. Increasingly this requirement has been explicated in terms of causal adequacy. Indeed, recent work on the method of "inference to the best explanation"[17] suggests that determining which among a set of competing possible explanations constitutes the best depends upon assessments of the causal powers of competing explanatory entities. Entities or events that have the capability to produce the evidence in question constitute better explanations of that evidence than those that do not.

Third, as the best explanation, B must have some comparative advantage over its principal rivals. Using hyperbole, we might say that it must be the *champion* among current competing explanations for A. B is therefore not the best explanation of A in any absolute sense. B must simply do a better job of explaining A than any of its current competitors. Explanation is therefore viewed as inherently competitive, contrastive, and fallible. Best explanations (champions) stand ever in need of critical reexamination. This third requirement therefore ensures that explanation is simultaneously progressive and self-critical.

Although this account gives only the barest sketch of what it means for B to be the best explanation of A, it will suffice for our purposes. Moreover, it accurately summarizes the development of Peirce's thinking in the hands of his modern-day successors. It is interesting to note that these modern-day successors are almost entirely philosophers of science. Imre Lakatos,[18] with his notions of competing "research programmes" and "heuristic power", Nancey Murphy,[19] with her application of Lakatosian philosophy of science to theology, Larry Laudan,[20] with his notions of competing "research traditions" and "problem-solving ability", and Peter Lipton,[21] with his carefully nuanced notion of "inference to the best explanation", all incorporate the basic criteria we have enumerated in their programs for scientific rationality.

How does epistemic support look when explanatory power rather than rational compulsion serves as its basis? The answer will by now be obvious. Instead of A epistemically supporting B because A rationally compels the acceptance of B, A now epistemically supports B because B serves as the best currently available explanation of A. And this in turn means that B is consonant with A, a contributor to our understanding of A, and the current champion among competing explanations of A.

The Big Bang and the Divine Creation

With explanatory power rather than rational compulsion characterizing epistemic support, the cosmological theory of the Big Bang and the Christian doctrine of divine Creation can now be brought into a relation of mutual epistemic support. To show this in detail far exceeds the scope of this modest essay. Still, a few brief observations will suggest how the Big Bang and the divine Creation might provide epistemic support for each other, once epistemic support is reconceptualized by reference to the logic of explanation.

Curiously, in the very passage in which he denies that relations of epistemic support obtain between the Big Bang model and the Christian doctrine of Creation, Ernan McMullin actually opens the door to such relations. In a passage already quoted, McMullin remarks, "What one could say . . . is that if the universe began in time through the act of a Creator, from our vantage point it would look something like the Big Bang that cosmologists are talking about. What one cannot say is, first, that the Christian doctrine of Creation 'supports' the Big Bang model, or, second, that the Big Bang model 'supports' the Christian doctrine of Creation."[22] Yet if we take explanatory power as our basis for epistemic support, it seems that what McMullin denies in the second part of this quotation he actually affirms in the first.

For consider what it means to say, "If the universe began in time through the act of a Creator, from our vantage point it would look something like the Big Bang that cosmologists are talking about."[23] Does this not simply mean that if we assume the Christian doctrine of Creation as a kind of metaphysical hypothesis, then the Big Bang is the kind of cosmological theory we have reason to expect? Does this not also mean that the Christian doctrine of Creation is consonant with the Big Bang? We submit that the answer is yes to both questions.

Suppose now that we take the Big Bang as given (= data) and pose the question of how we might best explain the Big

Bang in metaphysical terms. The playing field is potentially quite large. Metaphysics offers a multitude of competing explanations for the nature and origin of the material universe, everything from solipsism to idealism to naturalism to theism. Nevertheless, in practice we tend to consider only the competing explanations advocated by parties in a dispute. Since McMullin's foil is the scientific naturalist, let us limit the competition to Christian theism and scientific naturalism.

If we limit our attention to these two choices, Christian theism and its doctrine of Creation may with some justification be regarded as providing a more causally adequate explanation of the Big Bang than any of the explanations offered to date by scientific naturalism. Since the naturalist assumes that, in Carl Sagan's formulation, "the Cosmos is all that is, or ever was or ever will be",[24] naturalism denies the existence of any entity with the causal powers capable of explaining the origin of the universe as a whole. Since the Big Bang (in conjunction with general relativity) implies a singular beginning for matter, space, time, and energy,[25] it follows that any entity capable of explaining this singular event must transcend these four dimensions or domains. Insofar as God as conceived by Christian theism possesses precisely such transcendent causal powers, theism provides a better explanation than naturalism for the putative singularity affirmed by the Big Bang cosmology.

This assessment will no doubt seem unacceptable to the inveterate naturalist. And, indeed, many ingenious naturalistic cosmologies have been devised to circumvent both the Big Bang singularity and its apparent metaphysical implications. To see this one needs only to recall the contortions scientists have endured, not only in their metaphysical speculations but in their scientific theorizing, to avoid the dissonance created by the Big Bang cosmology for a naturalistic world view. Einstein acknowledged this dissonance when he introduced his notorious cosmological constant to maintain a static universe —a decision he came to regret, calling it the biggest blunder of his career. Fred Hoyle acknowledged it when he proposed his steady state theory to retain an eternal universe—despite its flagrant violation of the conservation of energy. Of course, most committed naturalists now reject both these theoretical formulations. And many would also acknowledge that a rudimentary logic of explanation does create dissonance between the Big Bang and naturalism. Nevertheless, they would assert that coupling Big Bang cosmology with more speculative quantum cosmologies or many-worlds hypotheses can eliminate dissonance. Yet, ironically, to the extent that even these cosmological ideas have validity, they may themselves have latent theistic implications.[26]

In any case, the Christian doctrine of Creation is consonant with a more standard Big Bang model and may well be regarded as a better explanation of it than its naturalistic competitors. Moreover, because the Big Bang is a putative scientific fact and because we are asking for a metaphysical account of that fact, it follows that the Christian doctrine of Creation is not a superfluous addition to our understanding of the Big Bang. The Christian doctrine of Creation actually contributes to our metaphysical understanding of the Big Bang by providing a causal explanation of it. Therefore, because Christian theism satisfies the first two criteria of best explanations enumerated above, it may (in a competition with naturalism) plausibly be regarded as a better explanation of the Big Bang. Hence, if we explicate epistemic support in terms of explanatory power rather than rational compulsion, it follows that the Big Bang provides epistemic supports for Christian theism and its doctrine of Creation.

To be sure, the argument that the Big Bang provides epistemic support for the Christian doctrine of Creation can be more fully developed and nuanced. Still, the general idea of how a fruitful interdisciplinary dialogue between theology and science may proceed should be clear. Note that in the example involving the Big Bang and the Christian doctrine of Creation, we only examined the case of a scientific claim (that is, the Big

Bang) providing epistemic support for a theological claim (the Christian doctrine of Creation). We could, of course, turn this around. Thus, we could fix the Christian doctrine of Creation as data and ask which cosmological theory of the origin of the universe is best supported by the Christian doctrine of Creation. The answer to this question is left as an exercise to the reader.

NOTES

[1] Rudolf Carnap, "Carnap's Intellectual Autobiography", in *The Philosophy of Rudolf Carnap*, ed. Paul A. Schilpp (LaSalle, Ill.: Open Court, 1963), p. 11.

[2] Ernan McMullin, "How Should Cosmology Relate to Theology?" in *The Sciences and Theology in the Twentieth Century*, ed. Arthur R. Peacocke (Notre Dame, Ind.: University of Notre Dame Press, 1981), p. 39.

[3] Cf. Wittgenstein's remark: "I am sitting with a philosopher in the garden; he says again and again, 'I know that that's a tree,' pointing to a tree that is near us. Someone else arrives and hears this, and I tell him: 'This fellow isn't insane. We are only doing philosophy'" (Ludwig Wittgenstein, *On Certainty* (New York: Harper and Row, 1969), p. 61e, no. 467).

[4] Here we are thinking especially of the work of Imre Lakatos, "Falsification and the Methodology of Scientific Research Programmes", in *Criticism and the Growth of Knowledge*, ed. Imre Lakatos and Alan Musgrave (Cambridge: Cambridge University Press, 1970), pp. 91–196; Larry Laudan, *Progress and Its Problems: Towards a Theory of Scientific Growth* (Berkeley, Calif.: University of California Press, 1977); Nancey Murphy, *Theology in the Age of Scientific Reasoning* (Ithaca, N.Y.: Cornell University Press, 1990); and Peter Lipton, *Inference to the Best Explanation* (London: Routledge, 1991).

[5] A detailed treatment of partial entailment may be found in Ernest W. Adams, *The Logic of Conditionals* (Dordrecht: Reidel, 1975).

[6] Charles S. Peirce, "The Red and the Black" (1878), in *The World of Mathematics*, ed. J. R. Newman, 4 vols. (Redmond, Wash.: Tempus, 1988), pp. 1313–14.

[7] Things become more complicated if in addition to probabilities we introduce utilities and thus have to balance the utility associated with a consequence against its probability (see Richard Jeffrey, *The Logic of Decision*, 2d ed. [Chicago: University of Chicago Press, 1983], chap. 1). Our discussion ignores utilities and focuses strictly on probabilities.

[8] This section summarizes the second author's treatment of explanation in his doctoral dissertation (Stephen C. Meyer, *Of Clues and Causes: A Methodological Interpretation of Origin of Life Studies* [diss., University of Cambridge, 1990]).

[9] Imre Lakatos, for instance, uses the phrase "heuristic power", whereas Larry Laudan speaks of "problem solving ability". See Lakatos, "Falsification", and Laudan, *Progress*.

[10] Charles S. Peirce, *Collected Papers*, ed. C. Hartshorne and P. Weiss (Cambridge: Harvard University Press, 1931), 2:372–88.

[11] Meyer, *Clues*, p. 25.

233

[12] Peirce, *Papers*, 2:375.

[13] Ibid.

[14] Ibid., 2:374.

[15] Synonyms and close relatives for consonance abound in the philosophical literature. These include *coherence, consistency,* and *consilience* (to name just a few that begin with the letter *c*). We prefer *consonance*, in part because it evokes the psychological notion of cognitive dissonance. Among theologians concerned with theology-science interconnections, *consonance* seems to be gaining ground in recent days (see Ted Peters, ed., *Cosmos as Creation: Theology and Science in Consonance* [Nashville, Tenn.: Abingdon, 1989]).

[16] See Lipton, *Inference*, pp. 114–22, as well as John Leslie's notion of "neat explanations" (*Universes* [London: Routledge, 1989]).

[17] Lipton, *Inference*.

[18] Lakatos, "Falsification".

[19] Murphy, *Theology*.

[20] Laudan, *Progress*.

[21] Lipton, *Inference*.

[22] McMullin, "Cosmology", p. 39.

[23] Ibid.

[24] Carl Sagan, *Cosmos* (New York: Random House, 1980), p. 4.

[25] Stephen Hawking and Roger Penrose, "The Singularities of Gravitational Collapse and Cosmology", *Proceedings of the Royal Society of London*, series A, 314 (1970): 529–48.

[26] Jay Wesley Richards, "Many Worlds Hypotheses: A Naturalistic Alternative to Design", *Perspectives on Science and Christian Belief* 49, no. 4 (1997): 224–26; Alvin Plantinga, *The Nature of Necessity* (Oxford: Clarendon Press, 1980), pp. 213–17; William Lane Craig, "Barrow and Tipler on the Anthropic Principle v. Divine Design", *British Journal for the Philosophy of Science* 38 (1988): 389–95; and William Lane Craig, "Cosmos and Creator", *Origins & Design* 17, no. 2 (1996): 26–27.